Farm Machinery Maintenance

NEWTON RIGG COLLEGE PENRITH

This book is due for return on or before the last date shown below

PRACTICAL FARMING

Farm Machinery Maintenance

PETER WHILEY

INKATA PRESS

INKATA PRESS
A division of Butterworth-Heinemann Australia

Australia
Butterworth-Heinemann, 18 Salmon Street, Port Melbourne, 3207

Singapore
Butterworth-Heinemann Asia

United Kingdom
Butterworth-Heinemann Ltd, Oxford

USA
Butterworth-Heinemann, Newton

National Library of Australia Cataloguing-in-Publication entry

Whiley, Peter.
Farm machinery maintenance.

Includes index.
ISBN 0 7506 8940 4.

1. Agricultural machinery – Maintenance and repair.
I. Title. (Series : Practical farming).

629.2875

Typeset by Ian MacArthur, Hornsby Heights, NSW.

Printed in Australia by Ligare Pty Ltd, Riverwood, NSW.

Contents

Introduction

Preventative maintenance is one of the most important tasks that can be performed on any machine. Time spent in the off season attending to these maintenance tasks will be well rewarded in the down time and expense saved during the working season.

Many of these maintenance tasks can be carried out in the farm workshop, if the operator has some knowledge and a few basic tools.

This manual aims to give farmers, machinery operators and students an insight into some of the problems and pitfalls that will befall any farm machinery operator. It gives step-by-step instruction on some of the service and maintenance tasks required to keep machinery in good working order.

To get the most out of any machine it must also be serviced regularly. This regular service includes checking various oil levels, greasing bearings, tightening chains and belts, and while performing all these service tasks looking for faults and defects on the machine. Any of these faults could cause trouble in the day's operation.

Because of the variety and range of machinery on most farms, the instructions are limited to specific components, like the fuel system, hydraulics or lubrication, rather than maintenance on a complete machine.

By servicing selected components, enough experience should be gained over a period of time to give the operator the knowledge and confidence to service a complete machine.

Note: always consult the operator's manual whenever operating or servicing a machine for the first time. It will give specific instructions on that particular machine.

The importance of safe work practice and safety in the workplace cannot be over-emphasized.

Safety

Precautions to be taken when undertaking any service or maintenance task:

1. Do not attempt any repair or maintenance task while machine is in motion or operating.

2. Do not operate engines in enclosed areas as this can cause CARBON MONOXIDE poisoning.

3. Do not clean components with petrol or flammable solvents without taking appropriate FIRE PRECAUTIONS.

4. Wear eye protection when using power tools and compressed air.

5. Wear a face shield when using pressure or steam cleaners.

6. Beware of hot components (exhaust, cooling system).

7. Beware of rotating components (fan, drive belts, pulleys, chains, sprockets fly wheels, PTO shafts). Ensure all safety guards are fitted correctly.

8. Disconnect the battery before major repairs or service to prevent accidental starting.

9. Do not cause spark near the battery (hydrogen gas may explode).

10. Select the correct tool for the task to be performed.

11. Ensure machine is on firm ground if jacking is required and ensure correct jacking points are used. Use hardwood blocks or suitable stands to support machine and chock all wheels not raised off ground.

12. Block all holes and passages into components with a clean rag or paper as parts are removed so small parts or dirt can not enter.

CHAPTER 1

Cleaners and solvents

Before attempting many of the service or maintenance tasks, the machine or component should be cleaned. Dirt and dust particles are one of the major causes of wear and breakdown in any machine. The importance of keeping every particle of dirt and dust out of machinery components when carrying out servicing, maintenance or repairs can not be over-emphasised. Operators should be ever vigilant to see that dirt does not get into any part of an engine, or machine working parts like the engine lubricating and fuel systems, bearings, gearboxes or hydraulic system. Dirt or dust between any moving parts will cause that component to wear much quicker than it would otherwise. The best way to stop dirt becoming a problem during service or maintenance is to remove it before any parts have been removed or dismantled.

Cleaning methods will vary with the machine or part being maintained or repaired. Grease nipples simply need a wipe with a clean rag before using a grease gun to force grease into the bearing to be lubricated. But it is still vitally important to remove any dirty grease or dirt and dust from that grease nipple before connecting the grease gun, as any dirt or dust on the tip of the grease nipple would also be forced into the bearing with the new grease, causing premature failure of that bearing.

This chapter discusses the various methods of cleaning machines and components:

- Cleaners and solvents
- Compressed air
- High pressure water (hot and cold)
- Abrasive blasting (sand blasting)
- Steam cleaners.

Cleaners and solvents

Solvents and detergents are very convenient for small parts that can be immersed in the solution. Solvent can also be sprayed or painted on larger components to help dissolve or soften hardened grease deposits before using a high pressure water cleaner.

Petroleum derived solvents are sometimes more convenient because of their availability. Kerosene and distillate are probably the best to use, because of their lower flash points. Only use petrol as a last resort, because of the danger of fire or explosion.

If using any petroleum based solvent as a cleaning agent make sure there are no sparks or naked flames that could ignite the solution. Some solvents may also have a detrimental effect on rubber, some plastics and electrical fittings.

Care must be taken when using any type of solvent, as they can cause skin irritation. Some will cause burning of exposed skin, and if splashed into the eyes can cause severe eye damage. Always read instructions and safety directions before using any chemical

---

After using any type of solvent always hose off the component with water and if necessary dry with a blast of compressed air. Never use compressed air above 100 psi, and never direct the blast of air at any part of the body especially the eyes.

Always wear eye protection when cleaning with compressed air

Air compressors

An air compressor has many uses in and around a farm workshop. They are not only used for inflating tyres, they can be used to clean components of foreign matter that can be simply blown away using an air director nozzle. Other uses for air compressors include, spray painting, all air tools, and many other tasks around the farm workshop.

Air tools available include:

- High and low pressure spray guns for painting anything from the sheep yard gates to the family car.
- Impact wrenches (rattle gun) and ratchet wrenches, invaluable for many jobs involving unscrewing and tightening nuts and bolts, like changing points on cultivators, chisel ploughs and combines.
- Air hammers, with punches, cold chisel, a miniature jack hammer. These have many uses in the farm workshop.
- Air drills and air screwdrivers and many other labour saving devices.

There are many types and sizes available, the size and type would depend on the use to which it would be put, but one with a capacity of 10 to 12 cubic feet (280 to 341 litres) per minute should be sufficient for most farm requirements.

Air compressors can be powered by electric motors or small petrol motors. Choice will depend on the availability of electricity, or of a generator if used away from the power supply.

Figure 1.1 *An Australian made air compressor suitable for farm workshop use*

Figure 1.2 *Cleaning a tractor radiator core with compressed air. Note the eye protection*

Ensure the one you choose is fitted with a pressure switch, pressure gauge, safety valve and a water and oil trap to re-move condensation and oil which may be forced past the rings on the piston. This ensures a clean air supply to tyres, spray gun and other equipment. If the compressor is also used for spray painting it will need a pressure regulator. This prevents damage to the spray gun by regulating the pressure supplied to the gun.

High pressure cleaner

High pressure water cleaners are very handy pieces of equip-ment to have in any farm workshop. They are available with electric or petrol engines. Pressure cleaners range from the small "cheapie" costing a couple of hundred dollars, to the heavy duty industrial type. One with a minimum pressure from 1500 to 2000 psi would be the best suited to farm work-shop uses; these will have enough pressure to lift off grease and caked dirt.

Other options that are available include; a rotating brush, rotating jet, a "turbo head" which in many cases increases the efficiency of the water jet, or a detergent or caustic clean-ing agent injector which helps in removing caked on dirt and grease. It is often better to soak hardened grease with a good de-greaser for a few minutes before using the pressure cleaner.

Figure 1.3 *A pressure cleaner in the medium price range*

Cleaners that heat their own water by various means may remove grease more easily, but are usually more expensive to buy and run. Another option on some pressure cleaners is a sand blasting kit.

- Care must be taken when using high pressure cleaners near electrical fittings in or around the workshop, as electricity and moisture can be a deadly mix.
- Take care when cleaning around gear box housings, transmission housings, or anywhere a shaft comes out of a housing. There will be an oil seal to retain the oil and pressures above 1500 psi will damage some seals and cause oil leaks. The high pressure can also force water past the seal and leave water in the gear box. This water may stay undetected until a bearing or gear failure occurs.
- Never clean the external core of a radiator with a pressure cleaner. This can bend or flatten the cooling fins and reduce the cooling efficiency of the radiator.
- Care must also be taken around paint work, as the water jet can remove even a baked enamel finish.
- Never use a high pressure cleaner on wood, particularly soft wood. It lifts the grain and leaves the wood very rough.

Sand blasting

Sand blasting nowadays is in name only. The silicon dust in sand causes silicosis, so all dry sand blasting is now done with ground copper slag, or ground limestone.

There are two types of sand blasting:

1. Wet; where the sand is drawn into the high pressure water hose and forced out under pressure with the water jet. These are probably the most universal for farm use, the sand only being used to remove harder, baked on grease, rust etc. Care must be taken when using near paint work as the sand blast will remove even the hardest baked enamel finish.
2. Dry sand blasters use compressed air as a propellent for the sand. They are mainly used in industry for rust and grime removal before painting.

Both types will leave the finished surface slightly pitted from the impact of the sand particles.

Steam cleaners

Very few steam cleaners are used in farm workshops these days. Their job has been almost entirely taken over by the high pressure water cleaners, which are cheaper to buy and run, and are more convenient and easier to use.

Steam cleaners require a heat source to heat the water in a boiler to steaming temperature, and a special steam hose and hand lance, insulated against heat to protect the operator.

The heat source to heat the water in a steam cleaner is usually an electrically operated diesel furnace.

CHAPTER 2

Lubrication

The most important maintenance task on any machinery is lubrication. Probably the chief cause of machinery wear is improper or insufficient lubrication. Whether it is checking the oil in the engine or gear box, or the regular inspection and greasing of the front wheel bearings, the lubrication service and maintenance task should be carried out regularly and methodically.

> **Always store lubricants in a clean air tight container. This will keep dirt and moisture out**

When greasing a machine always carry:

- a clean rag to wipe the top of each grease nipple
- a small spanner to remove any grease nipple that is damaged or "blocked"
- a piece of fine wire to unblock that grease nipple
- spare grease nipples to replace any damaged ones.

Sealed bearings on machines have reduced the greasing time considerably. Nevertheless all moving parts on a machine should be checked regularly.

When servicing a machine daily try and get into a system of starting on one side and working right around the machine checking each bearing, bearing housing and bolts, V-belt, pulley, chain, sprocket and grease nipple on the way.

Lubrication qualities

To lubricate properly and be effective, a lubricant must have at least three qualities:

1. The ability to make the surface slippery

2. The ability to adhere to the surface
3. The ability to maintain a film between the rubbing surfaces.

Specific purpose lubricants are made with additional qualities for their particular jobs. For instance, engine oil must have the above qualities, as well as the following:

- Have the ability to carry heat away from engine "hot spots" to where it can be cooled.
- Be thin enough to lubricate when cold, and thick enough to lubricate when hot.
- Have the ability to make a gas-tight seal between piston rings, piston, and cylinder wall.
- Have the ability to collect and hold grit, carbon and sludge, and convey it to the oil filter where it will be removed from the oil and held.

Why change oil ?

- To remove grit, carbon and sludge.
- To avoid a build up of crankcase oil dilution.
- To remove water caused by condensation.

Lubrication systems

You will encounter many types of lubrication systems as you service farm machinery. Listed below are a few, with points to look for when servicing machines equipped with the various systems.

Figure 2.1 *The simple oil can is a common manual lubricating device*

Figure 2.2 *Manual lubrication is required in this situation*

Manual systems

By far the simplest system of lubrication is the manual system. Whether the lubricant is solid, semi-solid or liquid, it is applied by hand. Any container can be used for liquid lubricants, while a grease gun, brush or spray gun can be used for the others (Figure 2.1).

Manual lubrication must be carried out regularly and methodically, otherwise failure of the component will result.

Pressure systems

A pressure lubricating system requires a pump of some description to supply oil under pressure to all parts of the item being lubricated. Most modern engines use pressure lubrication. The sump is the reservoir and the amount of oil in the sump should be checked regularly.

It is essential that whenever the oil level in the sump is checked, the tractor or machine is on level ground, and the engine should have been stopped long enough to allow all the oil from around the engine to return to the sump. On some larger tractors, checking should be done with the engine running. Consult the operator's manual for correct procedure.

The operator's manual will also list the correct grades of oil to use in that engine, and the operating pressure at which the engine should run. Fitting an oil pressure gauge is the best

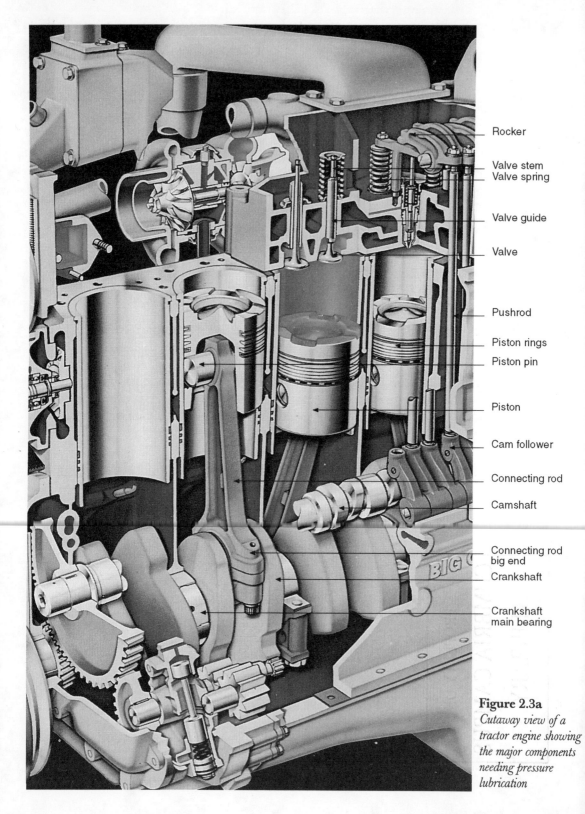

Rocker

Valve stem
Valve spring

Valve guide

Valve

Pushrod

Piston rings

Piston pin

Piston

Cam follower

Connecting rod

Camshaft

Connecting rod
big end

Crankshaft

Crankshaft
main bearing

Figure 2.3a
*Cutaway view of a
tractor engine showing
the major components
needing pressure
lubrication*

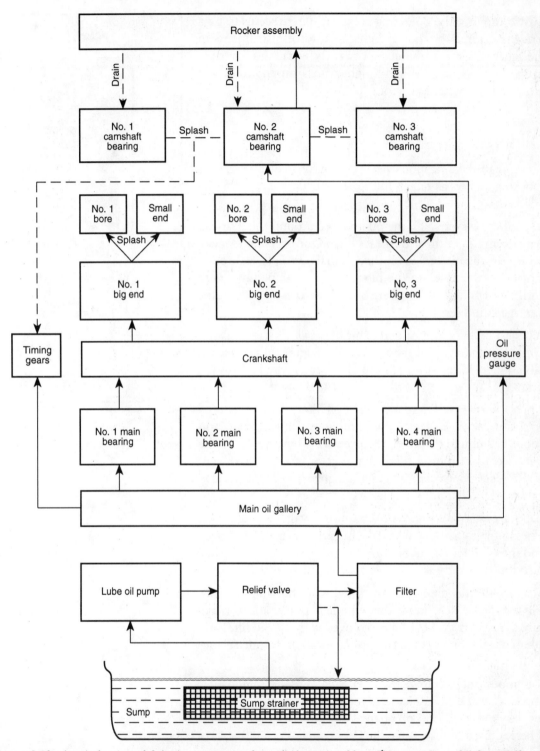

Figure 2.3b *A typical pressure lubrication system on a three cylinder engine. Note: This system has a full flow oil filter. If it was a bypass oil filter system, the oil filter would be located on the line returning to the sump from the relief valve.*

way to monitor the operating pressure. There seems to be a tendency to rely on "idiot" lights instead of instruments in the more modern machinery. For early warning of trouble gauges are essential. When a light comes on any idiot will know that something is wrong.

Oil filters

Most pressure lubrication systems will have an oil filter. These can be attached as either full flow or bypass filters (Figure 2.3b).

In a full flow system all the oil pumped around the lubricating system passes through the oil filter. One problem that can arise if the filter becomes blocked, is that oil will flow through a bypass valve fitted to the filter. The oil will be unfiiltered and may contain debris from the blocked filter, and some damage could result.

A bypass oil filter is one fitted to the bypass line, (Figure 2.3b). Only the oil that is not required by the engine, or any excess oil pumped, passes through the filter on its way back to the sump.

There are several different types of oil filters. The older type was a series of felt rings, approximately 12 mm thick and about 100 mm in diameter, (Figure 2.4). This type had to be removed from the oil filter chamber, washed in a solvent and replaced into the chamber. A great job for a Sunday morning!

Later types were usually of the throw away cartridge type. The oil filter chamber was detached, the cartridge removed and the oil filter chamber washed out. A new cartridge was then inserted and the chamber replaced (Figure 2.5).

Most modern engines are now fitted with the "spin on" cartridge type. When fitting this type it is important to make sure the rubber gasket ring is removed with the old filter and a film of oil is smeared on the gasket of the new one (Figure 2.6).

It is important to change the filter each time the engine oil is changed. Some manufacturers recommend changing the filter only every second oil change, but this is false economy. There is dirt and dirty oil in the filter container, and this dirty oil immediately contaminates the new oil when the engine is started.

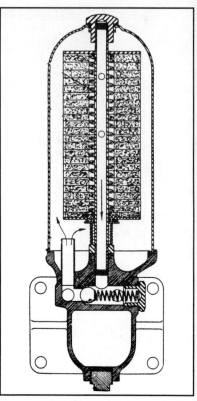

Figure 2.4 *Old type felt ring washable oil filter*

Figure 2.5 *Replaceable element type oil filter and rubber seal*

> **An oil filter will not remove oil dilution from unburnt fuel, water condensation and so on. Oil that turns dark quickly is doing its job**

Oil baths

This is the lubrication system used in most gear boxes. The oil level in the gear box covers the lower portion of some of the gears, and as the gears revolve they carry oil to the other gears, shafts and bearings in the gear box.

Care should be taken when adding oil to this type of system. If overfilling occurs extra pressure will be placed on the seals retaining the oil in the gear box. If too much pressure comes to bear on the seal it could fail and oil will leak out. Many gear boxes and differentials have to be dismantled to replace a seal.

Splash systems

Many of the small stationary engines and nearly all old engines use this type of system in the crankcase. They are usually fitted with a "dipper" on the big end bearing. As the crankshaft

Figure 2.6 *Spin on type oil filter and special oil filter removing tool*

revolves the dipper hits the oil and splashes it around the crankcase. This system is fast becoming obsolete in favour of the pressure system.

When using machinery fitted with small engines using the splash system it is important to remember to keep them level so the oil is where the dipper will contact it. The splash lubricating system was the downfall of many of the old cars and tractors that were used in hilly country. As they laboured up hills the oil ran to the back of the sump thus leaving the dipper on the front piston crank high and dry and restricting the amount of oil it could get.

Drip systems

Drip systems used to be a very popular system in old stationary engine designs, but more recently their use is limited to lubricating some chain drives, and some slow moving plungers. They are very simple in operation, having an adjustable orifice in the bottom of a container of oil, and allowing a variable supply of oil to drip onto the desired component (Figure 2.7).

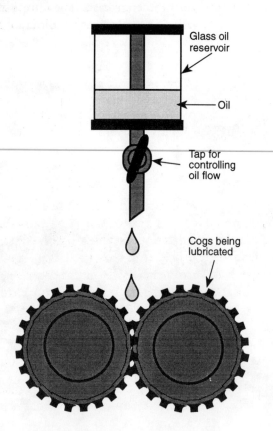

Glass oil reservoir

Oil

Tap for controlling oil flow

Cogs being lubricated

Figure 2.7 *Adjustable flow, drip system lubricator*

Types of oils and greases

There are many types of oil and grease available today. All should have an Australian Standard number in their specifications. This number should be checked with the machine or component manufacturer's recommendations to ensure the lubrication is suitable.

There are special oils for petrol engines, diesel engines, turbo enginges, LP gas engines, two stroke engines, automatic transmission gear boxes, hydraulics and differentials.

There are multi-purpose oils which combine characteristics of many of the different oils into one, but it must be remembered that each application requires a particular quality. To combine all the different attributes required into one oil must in some way decrease the quality of that oil.

Many engine oils are available in what is known as "multi-grade". This means they take on the character of different grades of oil at different temperatures. For instance a 20/40 oil is as thin or viscous as an SAE 20 oil when the engine is cold, which helps with cold starting and gives better lubrication to a cold engine. But as the engine temperature rises it takes on the thickness of an SAE 40 grade oil. SAE numbers followed by a W, (eg 10W/30) indicate oils that are intended for winter use.

Gear oils

Most manufacturers are now avoiding the use of thick gear oils and designing gear boxes to be lubricated with the thinner, more viscous, more refined gear oils. Modern additives give these oils the adhesion required to keep a film of oil between the gear teeth even under high loading.

Gear boxes are quite often located in inaccessible places and very hard to top up or fill. A gear oil pump is invaluable in these situations. If a pump is not available a simple way to get to that inaccessible gear box, is to pour the oil into a plastic drench backpack and use the hose without the drench gun to fill the gearbox. Squeezing the backpack will make the oil flow faster.

Be sure not to overfill the gearbox.

Hydraulic oils

Some oils are designated hydraulic oil. These are specifically for machines fitted with a separate hydraulic system, like most headers and fork lift trucks, some excavators and loaders. This special hydraulic oil has limited lubricating quality, is usually very thin and has a very high anti-foaming quality.

Oil used in the transmission of most agricultural tractors is the same oil used in the hydraulic system. The transmission housing acts as the hydraulic reservoir, so as well as being thin and anti-foaming, it also has to have very good lubricating quality for the transmission gears and bearings.

Some tractors use a manufacturer's specific oil in the transmission. If any other oil is used severe damage can result in the transmission and the hydraulic system. (See Chapter 7, Breakaway couplings.)

Always check the operator's manual for the correct oil to use in any machine.

Several machinery companies offer a service in oil analysis. Used engine, transmission or hydraulic oil can be analysed to detect contamination of any kind. The contamination can then be identified and traced to specific components, such as metal particles from bearings, or water from condensation, or a leaking gasket may let coolant into the oil. The analysis can even detect if the oil has been overheating. This service helps in diagnosing a problem that, as yet, has not been noticed in the field, because the machine is still working very well.

Greases

Grease is used as a lubricant where it is unsuitable to use oil, for instance where liquid oil would not stay in bearings long enough.

It is of utmost importance that different types of grease not be mixed. Grease is an oil or mixture of oils and additives, retained in a base substance. There are several base substances in use. The most common are lithium soap base and bentone clay base. These two types of grease are not compatible. If used together the oil and base material may separate, with the chance of leaving the bearings coated in bentone or bentonite, which is a fine clay. Clay on its own is definitely not a lubricant.

Many types of grease are designed for a specific purpose, check the operator's manual for suitable grease.

There are some very good multi-purpose greases available today. These greases are suitable for most greasing jobs in farm machinery maintenance, including plain, ball and roller bearings, universal joints and wheel bearings. These multi-purpose greases usually have a lithium base.

In high temperature conditions, such as front wheel bearings where disc brakes are fitted, a non-melting type grease must be used. Non-melting, high temperature greases are usually bentone based.

Other special purpose greases include:

1. Semi-fluid grease suitable for track rollers on crawler tractors.
2. Open gear lubricant for greasing open gears and flexible steel cables. It is a very good turntable lubricant on trailers.
3. Rubber grease; castor oil calcium soap thickened grease for use in contact with both natural and synthetic rubber. It can be used to pack rubber dust covers on hydraulic and mechanical brakes. If applied to wheel rims it will ensure easy tyre removal.
4. Anti-seize grease is a copper impregnated non-lubricating anti-corrosive grease. It was first developed for the drill rods on water and oil boring rigs. It stops the threads of the rods seizing, making disconnecting much easier. It is a grease that should be in every farm workshop. Anti-seize grease should be used in any situation where nuts, studs, pulleys, sprockets or bearings are likely to seize, rust or corrode onto bolts or shafts. A light smear on threads or shafts makes dismantling much easier.

Anti-seize grease is also ideal to use on exhaust pipe flanges, muffler connection, manifold studs, shafts and threaded fastenings, which might be exposed to chemicals and fertilizers.

Anti-seize grease is available under different trade names from various oil companies:

• Ampol make "Kopr-Kote Grease", available in 2.5 kg containers.
• BP make "Energrease AS1", also available in 2.5 kg containers.
• Shell make "Malleus TC2", but this is only available in 20 kg containers, probably too large for the average farm workshop.

Grease guns

Many types of grease guns are now available. The irksome task of filling a grease gun out of a larger container is all but passed. Grease guns are made to fit onto 2.5 kg or 20 kg containers and cartridges are made to slip into the smaller, hand grease guns. Be sure to remove the caps from both ends of the cartridge before slipping into the grease gun. Special pneumatic guns are also available to connect to an air compressor for pressure greasing.

The nozzle on the grease gun is fitted with three small jaws to grip onto a grease nipple. If these jaws or the grease nipple are damaged, or the grease gun is not held square to the grease nipple, the grease will squeeze out around the edges of the grease nipple.

Always try and organise to grease machines when the bearings are warm and the grease will flow much better.

Grease nipples

Grease nipples come in many different sizes, shapes and threads types. Figure 2.11 shows just a few. When replacing worn or damaged grease nipples make sure the thread type is the same on the replacement.

Dirt in many "plain bearings" tends to pack in the bottom of the grease nipple. To remove this compacted dirt, the grease nipple has to be removed, and the dirt dug out with wire until the shaft is visible. Do not use wire to remove the dirt from the base of the grease nipple; this may damage the small ball and retaining spring. Connect the nipple to a gun and force the dirt out with grease. The nipple should now be replaced and the bearing greased.

Greasing bearings

The method of grease application will often depend on the type of bearing being serviced (see also Chapter 4, Bearings and seals):

Plain bearings. Greasing should push out and replace old grease and dirt. Enough clean grease should be forced out of the bearing to remain as a "grease seal" to assist in keeping dirt out of the bearing. Do not leave grease around bearings where it could get onto parts like V-belts or clutch plates; it will make them slip.

Bearing with rubber dust caps. These include tie rod ends, ball joints and most universals. Grease until cap is full. Do not "pop"cap.

Sealed bearings. These are sometimes fitted with a grease nipple. The bearing should be filled with a grease

Figure 2.8 *Grease gun on a 2.5 kg container. (Courtesy Macnaught)*

Figure 2.9 *Grease gun on a 20 kg container. (Courtesy Macnaught)*

recommended by the manufacturer, at the recommended intervals. Do not overfill as this will damage the seal, letting dirt into the bearing reducing the life of the bearing.

Tapered roller bearings. Pack with correct lubricant as recommended by the manufacturer. Adjust to the specified preload if necessary.

> **Note**
> **Vehicles that travel on public roads should only be repaired by a qualified mechanic**

Figure 2.10 *Using a grease gun*

Figure 2.11 *A selection of grease nipples*

GJ 16	GJ 20	GJ 24	GJ 31	GJ 41	GJ 2	GJ 9
$\frac{5}{16}$" B.S.F. 35°	$\frac{5}{16}$" B.S.F. 67½°	$\frac{5}{16}$" B.S.F. 90°	$\frac{3}{8}$" B.S.F. Straight	$\frac{3}{8}$" B.S.F. 67½°	$\frac{1}{4}$" Whitworth Straight	$\frac{1}{4}$" Whitworth 67½°

GJ 13	GJ 21	GJ 50	GJ 58	GJ 29	GJ 34	GJ 39
$\frac{5}{16}$" Whitworth Straight	$\frac{5}{16}$" Whitworth 67½°	$\frac{1}{2}$" Whitworth Straight	$\frac{1}{2}$" Whitworth 67½°	$\frac{1}{8}$" Gas Straight	$\frac{1}{8}$" Gas 35°	$\frac{1}{8}$" Gas 67½°

GJ 44	GJ 30	GJ 48	GJ 52	GJ 56	GJ 60	GJ 77
$\frac{1}{8}$" Gas 90°	$\frac{1}{4}$" Gas Straight Short Thread	$\frac{1}{4}$" Gas Straight	$\frac{1}{4}$" Gas 35°	$\frac{1}{4}$" Gas 67½°	$\frac{1}{4}$" Gas 90°	Drive In $\frac{1}{4}$" Straight

CHAPTER 3

Fastening, locating and locking devices

Many components use some kind of fastener or locking device in their assembly. Fasteners can be defined as any device used to attach or secure two or more parts together. Locking devices are devices used to lock movable parts together. Locating devices are used to keep a part in its assigned place in relation to other parts in the machine. Some are relatively simple to remove, others need a particular tool or technique to remove them. These devices, tools and techniques are described in this chapter.

Threaded fasteners

Bolts and nuts are the most fundamental threaded devices used to hold two or more pieces of material together. They come in a range of sizes, thread types, left or right hand thread, head type, finish coating and tensile strength.

Until recently the most common thread type used on agricultural machinery was BSW (British Standard Whitworth) or Whitworth thread. The Whitworth thread has an angle of 55° and the top and bottom of the thread is rounded. Sizes are measured in imperial inches. They come in either black (mild steel with no other treatment) or cadmium plated finish, a shiny rustproof coating on a mild steel bolt.

High tensile bolts are usually supplied with a fine thread; BSF, SAE, UNF, metric etc.

Other thread types which are popular with Australian agricultural manufacturers include:

American
- SAE (Society of Automobile Engineers)
- UNF (United National Fine)
- UNC (United National Coarse).

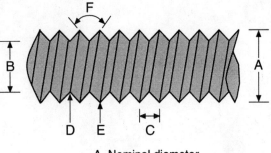

A Nominal diameter
B Root diameter
C Pitch
D Root
E Crest
F Pitch angle

Figure 3.1 *Detail of thread features giving a comparison of British and American threads*

British or **Whitworth** screw thread form has a round root (D), a round crest (E) and a pitch angle of 55°.

American National screw thread form has a flat root (D), a flat crest (E) and a pitch angle of 60°.

British

• BSF (British Standard Fine).

American threads differ from the British threads in pitch as well as thread angle and shape. American thread angle is 60° and the thread is flat on the top and the bottom.

American and British nuts and bolts come in imperial sizes.

Metric

Metric is the other bolt and thread type that is increasing in popularity since the Japanese entered the farm machinery market. The diameter of the bolt or stud, and the pitch, is measured in millimetres and the thread angle is 60°.

Metric bolts and threads have been in common use for decades in continental European countries involved with vehicle manufacture. Now with Japan manufacturing road vehicles as well as agricultural machinery, metric is set to be a universal type screw thread in Australia.

Bolt description

To order a nut and bolt you will need to state four things:

1. Bolt diameter

2. Bolt length

3. Head type

Figure 3.2 *Using a thread gauge*

4. Tensile strength. (High tensile bolts are about three times the price of mild steel. If only for ordinary work, mild steel would probably be strong enough.)

The thread type should not make any difference if bolts and nuts are bought together, provided you don't try to use different nuts or bolts when you get back to the workshop.

To order a stud, (a stud is a bolt that screws into a pre-tapped hole. It does not need a nut, the exception being wheel studs.) you will have to state the stud diameter, stud length and thread type. Studs usually have a hexagon head, unless used for a special purpose.

A thread gauge is invaluable in ascertaining the thread pitch (threads per inch) of a bolt.

Table 3.1 gives an indication of the difference between fine and course British threads, comparing them with UNC, UNF and metric. By studying the table it can be seen that the British coarse thread (Whitworth) and the American course thread (UNC) are almost the same, if comparing diameter and threads per inch. But it must be remembered that a Whitworth nut will not screw onto a UNC bolt, because the thread angle is different and the Whitworth thread is rounded at the top and the bottom whereas the UNC is flat.

The same rule applies to the British fine thread (BSF) and the American fine thread (UNF).

Most agricultural machines of British or American design or manufacture still use imperial sizes in bolts, nuts and studs.

Table 3.1 *Thread comparison chart*

Whitworth		British Standard Fine		UNC		UNF		Metric	
Diam. inches	*Threads per inch*	*Diam. inches*	*Threads per inch*	*Diam. inches*	*Threads per inch*	*Diam. inches*	*Threads per inch*	*Diam. mm*	*Pitch mm*
3/16	24	3/16	32	3/16	24	3/16	32	4.5	0.75
1/4	20	1/4	26	1/4	20	1/4	28	6	1.00
5/16	18	5/16	22	5/16	18	5/16	24	7	1.00
3/8	16	3/8	20	3/8	16	3/8	24	8	1.25
7/16	14	7/16	18	7/16	14	7/16	20	10	1.50
1/2	12	1/2	16	1/2	13	1/2	20	12	1.75
9/16	12	9/16	16	9/16	12	9/16	18	14	2.00
5/8	11	5/8	14	5/8	11	5/8	18	16	2.00
11/16	11	11/16	14	11/16	11	11/16	16	18	2.50
3/4	10	3/4	12	3/4	10	3/4	16	20	2.50
7/8	9	7/8	11	7/8	9	7/8	14	22	2.50
1	8	1	10	1	8	1	14	24	3.00

Left hand thread

Many of the nuts and bolts mentioned above are available with left hand thread, meaning that instead of turning the nut clockwise to tighten, it is turned anti-clockwise, and clockwise to loosen it.

This can be a very frustrating trap when trying to undo a tight nut. Machines with rotating parts held on with a nut sometimes use a left hand threaded nut and bolt that will work against the rotating part to keep the nut tight.

Some vehicles use left hand threaded wheel studs and nuts on the left hand side wheel. It can be very frustrating when trying to remove a wheel if you are not aware of this.

It must also be remembered that a left hand threaded nut will not screw onto a right hand threaded bolt or stud.

Bolt heads

Bolt heads can be any one of several types:

Hexagon head

Here a spanner is used to stop the bolt turning when tightening the nut. Nuts can be either square or hexagon, hexagons being the most popular, because ring spanners or sockets can be used with them.

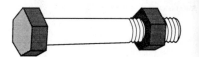

Figure 3.3 *Hexagon head, bolt and nut*

Cup head

The square under the "cup" on the head will fit into a recess or square hole in metal or will make its own recess when tightened in wood. This square is to stop the bolt turning when the nut is tightened.

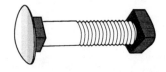

Figure 3.4 *Cup head, bolt with square nut*

Countersunk head

These are sometimes referred to as plough or carriage bolts. They usually have a square shank moulded into the countersunk head to stop the bolt turning when being tightened. They are used to bolt on plough shears, some cultivator feet, bailer plunger knife bolts, and anywhere a flat surface is required.

Figure 3.5 *Countersunk head bolt*

Cheese head or Socket head

These are shaped like the old block of cheese. They are used where a completely recessed head is required. They usually have a hexagonal socket recessed into the cylindrical head to take an Allen key for assembly purposes.

All the above bolt types are available in mild steel, or high tensile for special purposes. Many high tensile bolts are often

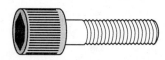

Figure 3.6 *Cheese head or socket head bolt with knurled head*

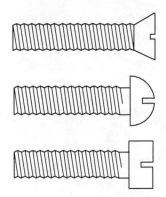

Figure 3.7 *Countersunk head screw, round head screw and cheese head screw*

Figure 3.8 *Thumb screw*

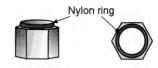

Figure 3.9 *Self-tapping screw*

Nylon ring

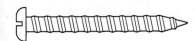

Figure 3.10 *Nyloc nut*

indicated by three dashes forming three triangles on the head of the bolt.

Machine screws

Machine screws come in the form of countersunk, round or cheese head crews. Countersunk head screws are usually used when joining two pieces of metal where one side has to have a flush surface. They are sometimes used with a nut on one side, or a threaded hole in the lower piece of metal.

Round head screws are used for the same purpose as above, but without the flush surface.

Cheese head screw are also used as above but the head usually fits into a recess in the top piece of metal.

Thumb screws

Thumb screws are used in many situations where only little force is required to hold a particular device. Some thumb screws have a hardened cup, others have a plain end.

Self-tapping screws

Self-tapping screws are designed to join sheet metal to sheet metal. A hole is drilled of the appropriate diameter for a particular size fastener, which is then screwed into the metal in much the same way as a wood screw is applied to wood. Self-tapping screws are available in a number of sizes and head patterns. They are usually steel and can be cadmium plated or stainless steel.

Wing nuts

Wing nuts are used on bolts that are not subjected to loads and allows the nut to be tightened and loosened frequently without the aid of a spanner.

Locking threaded fasteners

Nyloc self-locking nut

The Nyloc self-locking nut is a standard hexagon nut with a nylon insert slightly smaller than the diameter of the bolt. As the nut is tightened onto the bolt, the nylon insert binds onto the thread of the bolt and stops it from working loose.

Locking washers

Many types of locking washers are available; the situation will dictate which one will do the best job. All have the ability to bind onto the underside of a nut and stop it from working loose. It must be remembered that after several uses the ability of the washers to bind onto a nut is reduced, and they may have to be renewed. In critical situations where damage could result if a nut works loose, it is advisable to fit a new locking washer each time the item is dismantled.

Spring washers come into this category. They are used under nuts to stop the nut from working loose. They should be replaced after several uses, as the spring tension decreases and the sharp edge that grips the nut tends to wear out.

External tooth lock washers are also used to stop nuts from working loose. They have more gripping points than a spring washer.

Internal tooth lock washers do the same job as the external tooth washer.

Lugged washers are used in special situations. The lug is bent into a slot in a bearing or the flat side of a nut. Some internal lugs fit into keyways on shafts and axles.

Sometimes the locking washers can be dispensed with if Nyloc self-locking nuts are used where vibration tends to work ordinary nuts loose.

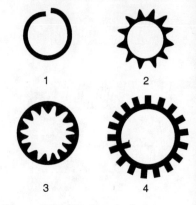

Figure 3.11 *Locking washers:*
1. Spring washer
2. External tooth washer
3. Internal tooth washer
4. Lugged washer

Spanners and sockets

It must be remembered that most spanners and sockets designed to fit British thread bolts will not correctly fit American or metric threaded bolts, and vice versa. However some can be used with caution. Metric spanners and sockets with 1 mm increments will probably fit the widest range of bolts and nuts, although they are not always a good match.

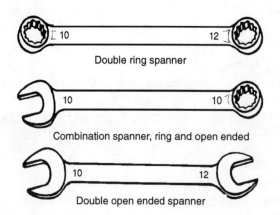

Figure 3.12 *Spanner types*

There is a chance of the spanner slipping and damaging the corners of the nut or bolt, (and removing skin from the knuckles of the operator).

Spanners and sockets come in different forms:

- Open ended
- Combination
- Ring spanners
- Sockets, with ratchet handle
- Box spanners.

Non threaded fasteners

Rivets

Rivets are made of a soft malleable mild steel. They are not nearly as popular for fixing as they used to be because of the high labour input required in a good riveting job, (the Sydney Harbour Bridge is all riveted together). Welding and spot welding have taken their place in many situations.

One application for mild steel rivets is attaching mower and header knife sections, and the knife head, to the knife cutter bar.

To replace a riveted knife section in a header or mower knife, remove the broken section. Hold the knife bar square on the top of a vice. One firm blow with a hammer should sever the old rivet.

To fix a new knife section, place a rivet in the knife bar and position the rivet over an anvil or some solid object. Place a new knife section over the rivet and burr it with a hammer just enough to stop the rivet falling out. Insert rivets in the other holes and repeat the last step. Continue to burr the rivets over with the ball peen of an engineer's hammer until

Figure 3.13 *Round and countersunk head rivets*

Figure 3.14 *Removing a knife section from a knife cutter bar*

the rivet is finished to a round head. The round head finish is important, because it will be much stronger than a finish that is just hammered flat. The knife section should now be tight and firm on the knife bar.

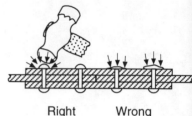

Right   Wrong

Figure 3.15 *Right and wrong methods of riveting*

Pop rivets

Pop rivets are very useful for fixing two sheets of sheet metal together, or fixing sheet metal to other thin or light metal frames. They come in various diameter sizes, length and types of metal. Aluminium alloy and mild steel are the most popular, although Monel metal (a high tensile nickel-copper alloy) and copper are also available. Pure aluminium pop rivets are also available for fixing hardened plywood, perspex, brick and so on.

A hole of the correct diameter to suit the pop rivet must be drilled through the metal. A pop rivet can then be placed in the hole, making sure that both pieces of metal to be joined are tight together. A special pop rivet gun is then placed over the "nail" and the gun operated until the nail breaks. The job is now complete.

Two main types of pop rivet fixing guns are available; hand plier type, suitable for smaller rivets up to 4 mm diameter and lazy tong or cantilever type, for rivets up to 5 mm diameter.

Pneumatic or hydraulic pop riveters are also available for pop rivets up to 6 mm diameter, but this type would probably be out of the range for farm workshop use.

Figure 3.16 *Rivetting tools;* left, *hand plier type and* right, *lazy tong type*

Locating devices

Keys

Keys are used in pulleys, sprockets and cogs, where more torque is required between the device and the shaft. They vary in size and shape, but all require a keyway to be machined into both the shaft and the pulley or sprocket.

Gib head key

This is a slightly tapered key that is driven into the matching keyways in the shaft and pulley. To remove the key, a drift (tapered metal tool) is driven between the key's protruding head and the pulley boss, forcing the key out along the keyway in the shaft.

Straight key

As its name suggests this is a straight key, without the head. It is usually held in place on the shaft and in the pulley with a set screw.

Woodruff or half moon key

These need a special half round keyway in the shaft. This keyway must be the same size as the key. The pulley is held in place over the key on the shaft with a set screw.

Pins and other locating devices

Sel-lock and roll pin

These pins can be used to replace keys and set screws when locating a pulley or sprocket on a shaft. The disadvantage is that if the pulley has to be realigned, a new hole will need to be drilled through the shaft. This may weaken the shaft.

The roll pin is driven into a hole slightly smaller than the diameter of the pin. The expanding pressure of the pin then holds it in place in the hole. In some larger size pins, an additional smaller pin may be driven into the centre of the larger pin. This gives the larger pin extra expanding pressure for heavy duty tasks.

A pin punch of the correct size is essential to remove roll pins. The correct size is smaller than the inside diameter of the hole through the pulley or sprocket, but larger than the diameter of the hole in the roll pin. If a smaller pin has been driven into a larger pin, it will be easier to remove the smaller pin first, and then the large pin. This will require two different sized pin punches.

Figure 3.17 *Gib head key*

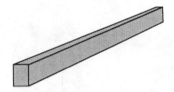

Figure 3.18 *Straight key*

Figure 3.19 *Woodruff key*

Figure 3.20 *Sel-lock pin*

Cotter pins

Cotter pins are used as locating and locking devices on many types of equipment. Who would have ever made a billy cart without having a cotter pin to hold the wheels on. Many times a piece of wire or a nail has substituted for a cotter pin in an emergency, but it must be stressed that these substitutes are only to be used in emergencies as they will not stand the pressures and wear that a cotter pin of the correct size will.

Cotter pins have many uses as locking or locating devices in shafting, bearings, pins or nuts, to stop the slotted nut in the front wheel bearing from working loose. In this application they must be fitted in the correct way. Always use a washer the correct size behind the cotter pin.

Figure 3.21 *Cotter pin*

Slotted and castle nuts

Slotted and castle nuts are used in conjunction with cotter pins usually where the nuts do not have to be tightened right up, as in the case of front wheel bearings, or where there is excessive vibration that could work the nuts loose. If replacing these nuts, diameter and thread type must be stated.

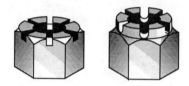

Figure 3.22 *Slotted and castle nuts*

Set screws

Various types of set screws are used to locate many turning devices like cogs, sprockets and pulleys, onto shafting.

The square head, case hardened cup point set screw is the most popular. The correct size spanner is the only tool required to loosen or remove this set screw. The hardened cup point bites into a shaft and helps in retaining and locking the attachment to the shaft.

Figure 3.23 *Always fit cotter pins in slotted or castle nuts as shown*

The hexagonal socket head screw (Allen grub screw) also has a hardened cup and is used for the same job as the square head set screw. It has the added advantage of having no protruding parts that could catch and tangle things around the revolving shaft. A special hexagonally socketed head key (Allen key) of the correct size is required to remove this type of set screw.

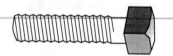

Figure 3.24 *Square head set screw. Note cup point*

It is as well to remember that the Allen keys come in metric and imperial sizes, and they are rarely interchangeable.

Circlips

Circlips are used in various situations usually to locate a bearing on a shaft or in a housing. They fit into a machined groove, either external or internal. Special circlip pliers are used to insert or remove them. Circlip pliers are made for internal or

Figure 3.25 *Hexagonal socket head screw*

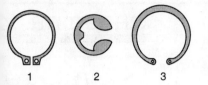

Figure 3.26
1. External circlip
2. 'E' type external circlip
3. Internal circlip

external circlips. When the handles of internal circlip pliers are squeezed together the pins that grip the circlip also come together, but when the handles of the external pliers are squeezed together the circlip pins are forced apart.

When removing circlips always put something over the clip in case it flips off the end of the pliers. They are very hard to find in a straw paddock or any where else for that matter!

Before attempting to remove any type of bearing from a shaft or housing you should determine how the bearing is fixed or located, and remove that locating or locking device first. It may save damage to the bearing or housing. Circlips are sometimes very hard to see in confined areas, so make sure all is clear before force is applied to bearings or housings.

CHAPTER 4

Bearings and seals

Inspecting and servicing bearings and seals is essential in a preventative maintenance programme. This chapter includes some causes of bearing failure in different types of bearing and seals, and how these failures might be prevented.

Plain bearings

Plain bearings are the simplest form of bearing. The revolving shaft or sliding surface is in direct contact with the fixed bearing surface. These metal to metal contact surfaces should always be of different metals, the bearing metal being softer than the shaft or slider. This allows the bearing to suffer wear, rather than the shaft or slider. Many plain bearings are "split", that is, made in two halves, and are designed to be relatively easy to replace when they are worn out.

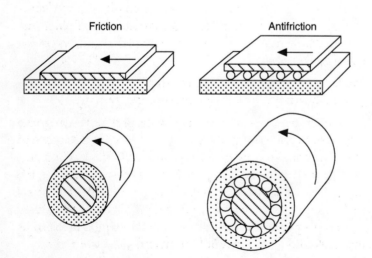

Figure 4.1 *Bearings can be classified as friction (sliding or plain bearings) or antifriction (rolling ball or roller bearings*

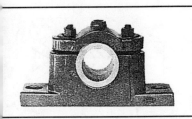

Figure 4.2 *Examples of plain bearings. Above, standard cast iron plummer block with gunmetal bearing and right, wooden straw walker crank bearings (courtesy Claas Headers)*

Another reason for using dissimilar metals in bearings is to reduce the likelihood of like metals binding. This occurs when two metals of the same composition, rubbing together as in a plain bearing, cause friction. One bearing surface tends to snag or bind on the other. It eventually leads to bearing seizure.

Common materials used in conjunction with steel shafting or sliders include white metal or babbitt alloys (one of the most popular), cast iron, lead-copper alloys, brass, bronze alloys, aluminium alloys, plastic and even wood.

Examples of plain bearings in agricultural machines include most big end and main bearings on engines. Many slow moving shafts have plain bearings, particularly on older machines. Many headers still use wooden plain bearings on the straw walker crank.

Metal to metal contact has a high friction factor. For this reason the bearing must be kept lubricated.

Most plain bearings are lubricated with grease. When greasing these bearings with a grease gun, the new grease should force out the old grease and dirt and leave the bearing surfaces coated with new grease. Some grease should be left around each side of the bearing. This grease will form a grease seal, helping to prevent dirt entering the bearing. In dusty conditions plain bearings can require lubrication as often as every two hours to prevent friction and excessive wear.

Engine bearings are pressure lubricated. The oil is circulated to each bearing and after being forced through the bearing finishes up back in the sump (see Lubrication, Figure 2.2). Other bearings are lubricated with a drip system where oil drips into the bearing from a controlled drip apparatus, (see Figure 2.7).

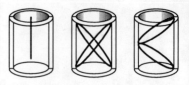

Figure 4.3 *Plain bearing bushings showing various lubrication groove patterns*

Lubrication plays a major role in reducing friction. If correct and thorough lubrication is not carried out, bearing failure and machine damage is a certainty. The aim of lubrication is to put a film of oil or grease between the two bearing surfaces and keep it there, so that the shaft is actually suspended on a film of oil or grease inside the bearing.

Dirt, dust and other foreign matter are other major causes of bearing failure. Dirt acts as a grinding paste in a bearing and will soon wear away the metal on the bearing surfaces, resulting in the bearing being a very loose fit and causing gears to mesh incorrectly, (see Backlash clearence, Figure 5.32), sprockets and chains to run out of alignment and the whole machine to run poorly or rattle badly.

Anti-friction bearings or rolling elements

The following bearings are classed as anti-friction bearings because they incorporate rolling elements such as balls or rollers, held between two raceways. Because of this rolling action there is considerably less friction than in the plain bearing.

Ball bearings

From the time of our first bicycle we were aware of ball bearings, and how much better the bike seemed to go after we oiled the wheels and the pedals. These parts are fitted with anti-friction ball bearings. After lubrication, friction is reduced even further.

Ball bearings come in many different types and sizes and they have many different applications. Most shafts on agricultural machines will be supported with a ball bearing of some kind. Headers, bailers, mowers and ploughs all use ball bearings.

Most ball bearings fitted to agricultural machinery are sealed, meaning that the bearing has a seal or cover on one or both sides to stop dirt, dust and water from getting into the inside of the bearing.

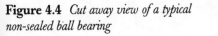

Figure 4.4 *Cut away view of a typical non-sealed ball bearing*

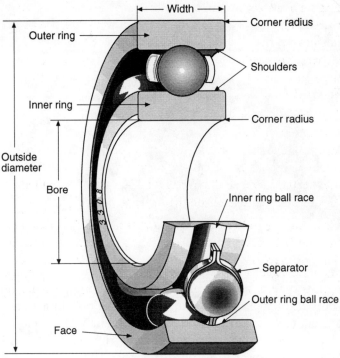

Single side seals are usually used on gear boxes. Oil in the gearbox is able to lubricate the balls or rollers in the bearing. The seal should stop dirt entering and oil escaping. In most applications like this there is usually a separate lip seal on the outside of the bearing (see Figures 4.21 and 4.24). If the seal becomes damaged in any way it will not be an effective barrier. Dirt will get into the gear box, mix with the lubricant and grind away at the bearing and it will soon fail.

Damage can be caused to seals by bailer twine wrapping around the shaft and bearing, resulting in distortion or splitting of the seal, or causing the seal to pop out completely. The same thing can happen with vegetable matter such as wire weed, straw and thistles. Saffron thistles in particular are very stringy and tend to wrap around shafting. Bearings on disc ploughs, slashers, headers and potato diggers are very prone to this sort of damage.

Many sealed bearings have a grease nipple. Care must be taken when greasing to only pack the bearing and not over-fill it and pop off the seal.

Disc ploughs fitted with sealed bearings and a grease nipple are usually open ball or roller bearings with a separate lip

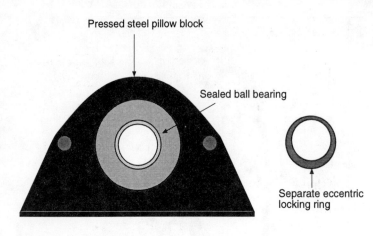

Pressed steel pillow block

Sealed ball bearing

Separate eccentric locking ring

Figure 4.5 *Sealed ball bearing in self-aligning pressed steel pillow block. Note: separate eccentric locking ring*

seal. This seal is to keep dirt out of the bearing. When greasing this type of bearing, grease should be pumped into the bearing, forcing any old and dirty grease out, in a similar fashion to plain bearings. When replacing lip seals on plough bearings, be sure to install them in the right direction, with the lip facing towards the outside.

Roller bearings

Roller bearings have hardened rollers instead of balls in the bearing. They have a high radial load (at right angles to shaft) capacity, but no end thrust (parallel to shaft) capacity. They are more sensitive to misalignment and dirt contamination than ball bearings.

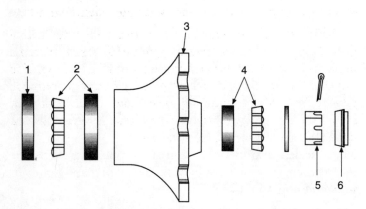

1. Seal
2. Inner tapered roller bearing and cup
3. Wheel hub
4. Outer tapered roller bearing and cup
5. Slotted nut and pin
6. Grease cap

Figure 4.6 *Typical wheel bearing with tapered roller bearings*

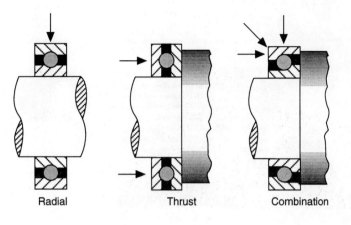

Figure 4.7 *Loading designs. Arrows indicate direction of load*

Radial   Thrust   Combination

One of the most common applications of a roller bearing is the tapered roller bearing. These have end thrust and radial load capacity and are found in most wheel bearings. They are adjustable and should be checked regularly for wear and movement in the bearings, particularly if the front wheels are subjected to heavy load.s. The tapered roller is made up of an inner ring the rollers held in a cage and the outer cup or cone.

Needle bearings

Needle bearings are a form of roller bearing using many more, but much smaller diameter rollers than a roller bearing. They are used where space is limited, as in most universal joints and some gear boxes. They may be constructed without a race for balls or rollers (universal joint bearings), with an outer race only, or with an inner and outer race (Figure 4.8).

Agricultural universal joint bearings are usually fitted with a grease nipple. A rubber grease cap on each leg of the cross prevents the entry of dirt and moisture into the bearing. The universal should be greased regularly with a multi-purpose grease. Always stop pumping grease into a universal joint before the cap pops off. Two or three pumps once daily are quite sufficient in most circumstances.

Bearing housings

Bearing housings are machined into gear boxes, machine frames and anywhere a shaft has to be supported by a bearing

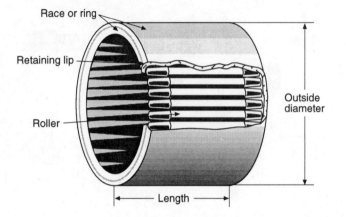

Figure 4.8 *Needle bearings with outer race only*

in a non-adjustable location. The machined housing always has parallel sides.

Pillow, plummer or pillar block

Cast or pressed metal housings are used as a bolt on bearing support for a shaft. They can be made to suit a plain bearing, parallel sided ball bearing, or self-aligning bearing, (see Figures 4.5. 4.9 and 4.10).

Self-aligning pillow block bearing housings are concave inside to take a convex shaped bearing ring. This type is called self-aligning because they can turn laterally in the bearing housing if the shaft and housing are not at a true 90°.

This type of bearing housing is used extensively in farm machinery where the shaft has to be moved for the adjustment of V-belts or chain drives. Very little extra strain is placed on the bearings if the shaft is not adjusted exactly square.

Self-aligning flangette bearing housings

These cast or pressed metal bearing housings do the same job as a pillow block bearing housing, differing only in that the bearing is bolted onto a flat surface, with the shaft going through that flat surface.

Locking devices in bearings

Most bearings have a locking device of some description to lock them on the shaft. This device is to:

1. Grease nipple, must line up with hole in outer ring of bearing
2. Grease hole in outer ring — under grease nipple
3. Outer seal — metal or hard plastic
4. Inner rubber seal — on inside of outer seal
5. Hexagon socket head set screw, to locate bearing on shaft
6. Cast iron one piece housing
7. Convex outer ring fitting the concave housing
8. Ball raceway
9. Wide bearing inner ring

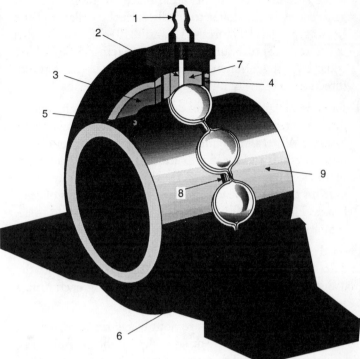

Figure 4.9 *Cast iron pillow block housing with bearing*

• stop the shaft turning inside the bearing inner ring,
• stop the shaft sliding laterally in the bearing.

There are several ways to lock the bearing on the shaft:

1. Hexagonal socket head screw (Allen grub screw). When placed through the inner ring of the bearing and tightened onto the shaft, this will hold the bearing or shaft in place, and stop the shaft turning in the bearing (Figure 4.10).

2. Tapered adaptor. This type of locking device relies on a nut tightening onto a tapered thread. The more the nut is tightened, the tighter the tapered bush is compressed onto the shaft. The nut must be locked with a locking washer by bending one of the lugs into the slot in the nut (Figure 4.11).

3. Eccentric locking collar. The eccentric collar is turned against another eccentric on the inner ring of the bearing. It is important when installing this type of fixing device that the eccentric outer locking collar is turned in the same direction as the shaft rotation. A hole is provided in the outer locking collar to allow it to be tightened with a hammer and punch. Give the punch a good sharp hit to

Grease nipple hole

Hexagonal socket head screw (Allen grub screw)

Figure 4.10 *Cast iron flangette bearing housing and bearing. Note: grease nipple hole, locking grub screw*

make sure the collar is locked in place. Tighten the Allen grub screw in the collar to finally lock the collar and the bearing in place (Figures 4.12 and 4.13). To remove this collar, reverse procedure. Remember to remove the locking collar by using a punch to rotate the collar in the opposite direction to the way it was tightened onto the shaft.

4. Circlips are used to locate a bearing on a shaft, with a groove machined into the shaft where the bearing is to be located. When assembling, the bearing is pressed or driven onto the shaft and the circlip inserted on one or both sides of the bearing, depending on the design of the component. Circlips are also used to locate bearings in housings of gearboxes. Here a groove is machined into the housing to accept an internal circlip.

Pressed steel bearing housings just come apart when unbolted and do not pose any problem to replace the bearing into the housing. In fact it is usual to fit a new housing each time a bearing is replaced.

Cast iron bearing housings are not always replaced when a new bearing is fitted. There are two slots in the bearing housing (Figure 4.14 top). The outer ring of the bearing is held at 90° to the bearing housing and slipped into these slots. The bearing is then rotated into the concave housing, (Figure 4.14 bottom), the bearing being brought up to the vertical position in the housing. Note: the grease hole in the outer ring of the bearing must line up with the grease nipple in the housing.

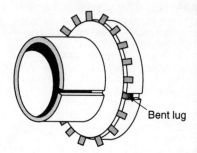

Bent lug

Figure 4.11 *Tapered adaptor with lugged locking ring*

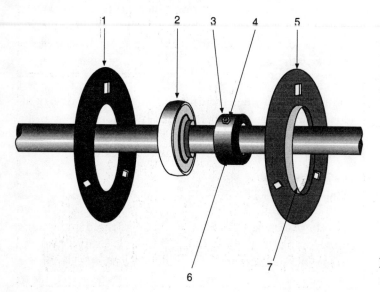

1. Bearing housing
2. Bearing
3. Eccentric collar
4. Grub screw
5. Bearing housing
6. Hole for punch
7. Extruded side of bearing housing

Figure 4.12 *Sealed flangette bearing with eccentric collar locking ring*

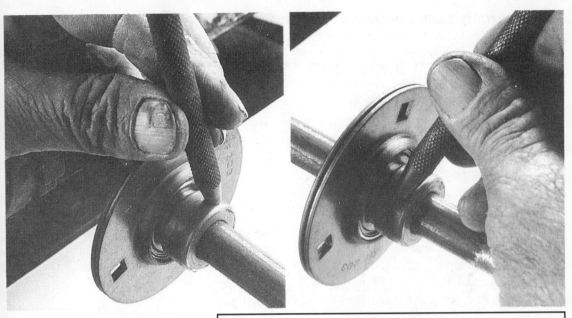

Figure 4.13 Left, *tightening collar on a sealed flangette bearing and* right, *loosening it (in both cases the shaft rotates in a clockwise direction)*

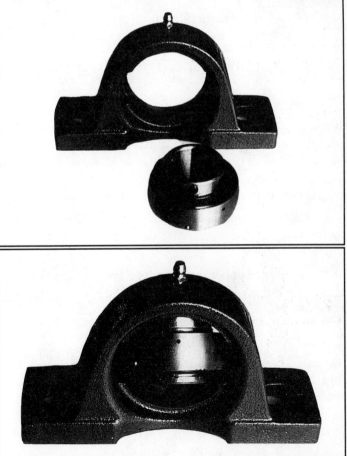

Figure 4.14 *Cast iron pillar block housing with,* top, *slots on either side of the bearing opening and* bottom, *the bearing being rotated into the housing*

General maintenance on anti-friction bearings

Always maintain absolute cleanliness when inspecting, adjusting, washing or installing bearings. Bearings should be kept free of dirt and dust at all times.

When cleaning bearings only use clean solvents. If dirt or grease is washed out of a bearing, rinse again with clean solvent. Wash solvent out of bearing with a low pressure water jet and use compressed air to dry the bearing. Remember to use protective clothing and a face mask when cleaning. Never use compressed air above 100 psi. Never spin a dry (un-lubricated) bearing.

The first indication of bearing failure will be the bearing getting hot. If a bearing becomes too hot to hold your hand on, it is a sign that all is not well. The heat will be caused by excessive friction between the moving parts in the bearing as it rotates. This is caused by:

- lack of, or insufficient lubrication
- dirt or dust in the bearing, entering through a damaged seal
- overloading of that bearing by excessive tension of chains or belts
- misalignment of non self-aligning bearings.

If the heat is caused by lack of lubrication or dirt several pumps of grease into the bearing may help, provided a grease nipple is fitted. If no grease nipple is fitted, sometimes oil can be injected into a sealed bearing using a hypodermic needle inserted through the neoprene seal of the bearing.

This is a very temporary measure and should only be used to finish a job. A new bearing should be fitted at the first opportunity.

When ball bearings become worn to the stage of rattling they should be replaced before more damage is caused to other components.

Tapered roller bearings: maintenance

Tapered roller bearings are commonly used on wheel bearings. They should be checked regularly for wear, removed, washed and inspected according to the machine manufacturer's recommendations. Tapered roller bearings are also used in some differentials, gear boxes and some plough discs, but these are not as easy to remove and check.

Tension wrench

Stand

Figure 4.15 *Tensioning a wheel bearing to the reccommended pre-load with a tension wrench*

To check wheel bearings for wear; chock remaining wheels to prevent machine from rolling. Take the weight off the wheel bearing to be checked. Once the wheel is off the ground always chock the implement with a solid block.

Check movement in the wheel bearing. Adjustment will be necessary if movement is excessive. If the bearing is left with excessive movement, damage to the bearings could result. The constant thumping of the wheel is likely to damage the nut and thread, even to the extent of stripping the thread completely. This could result in the wheel simply falling off. Threads on stub axles are difficult to repair.

To adjust, leave the wheel off the ground and chocks in place. Remove grease cap and locate locking device, (lock washer or cotter pin). Remove this device and tighten nut, while turning the wheel until it just begins to bind. Then loosen the nut one notch of the locking device (slotted nut or lugged locking washer).

If the manufacturer recommends a "pre-load", the nut must be tightened to this specification using a tension wrench, then backed off one notch of the locking device. Check that the wheel spins freely without binding. Replace grease cap and check other wheels.

To remove, wash and inspect tapered roller or wheel bearings it will be easier to remove wheel first. Then remove the grease cap and locking device as above. Remove nut, taking care not to let washer and bearing drop off the shaft. Always place a clean lint-free cloth or clean tray under work in case of slip up. The bearing should come out complete with roller race. If rollers fall out of the race it is an indication of excessive wear, and the bearing and ring should be replaced.

Carefully remove hub and there will be another tapered roller bearing on the inside of the stub axle. This bearing should pull off the axle. Take care not to damage the roller race or the seal behind the bearing. The two bearing rings will remain in the hub and can be inspected in place. Examine for any uneven marks on ring surface such as pitting, surface flaking, grooves or scratch marks; anything that is not an even shiny surface.

Wash the rollers and race in a good solvent to remove all old grease. Dry and inspect the rollers for wear and uneven surface. If any of these symptoms are present it indicates the bearing could fail in the near future. The bearing and matched ring should therefore be replaced.

Note: Vehicles that travel on public roads should only be repaired by a qualified mechanic

---

When installing an anti-friction bearing it is important that an even pressure is applied around the ring of the bearing. It will depend on the design of the bearing and housing which ring is an "interference fit" and has to be pressed or driven into place first. As a general rule the rotating ring is usually the one to be pressed or driven home, while the stationary ring is a slip or push fit. (The cup is a press or driven fit in a wheel hub, while the inside cup, the stationary one, is a push or slide fit on the stub axle.)

In most gear boxes the interference or drive fit is reversed. The shaft rotates so it has the driven or press fit. The gear box is stationary so the bearing is a slide fit into the machined gear box housing. It is usually held in place in the housing with a circlip.

Caution should be taken if an anti-friction bearing has to be pressed or driven into a cast iron housing in a gear box. The cast iron is often very brittle and if too much force is applied the casting will break.

It is not always convenient to use a press when fitting bearings. There are special mounting dollies and sleeves available for fitting bearings, but these are rarely found in a farm workshop. Below is the more common method of installation:

1. Select a clean work area free of dust, loose dirt and anything else that may get into bearing.

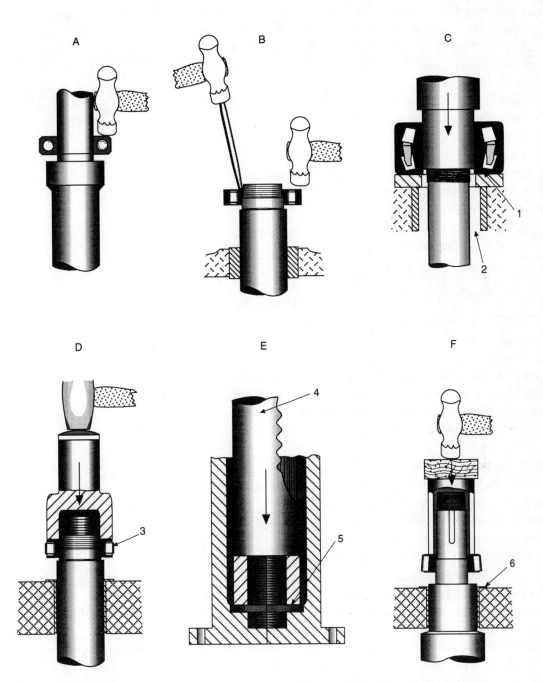

A. Do not strike the bearing with a hammer.
B. Do not use wide punches on bearings.
C. Apply force to tight ring (1) and have clearance (2) for the shaft.
D. Use a driver with smooth and square cut ends that strike the tight ring (3).
E. Force the ring to the full depth with a suitable pipe (4) into a cleaned bearing ring recess (5).
F. Block placed on open pipe driver allows driving force to be centralised. Use protective vise jaw covers (6).

Figure 4.16 *Bearing installation hints. (Courtesy Anti-Friction Bearing Manufacturers Association)*

2. Determine which cup or ring is the drive fit, and find a piece of steel pipe about 2 mm smaller in diameter if driving into a housing or 2 mm larger if driving onto a shaft. Make sure the pipe is cut square on the end to be applied to the bearing.

3. Clean the pipe of any loose dirt or rust and any burrs or loose metal on the end.

4. Grip the housing or shaft, whichever one has the driven fit bearing ring, firmly in a vice or suitable place. Take care not to damage either. It may be necessary to use vice jaw protectors, (soft sheet metal covers over the vice jaws) to protect the parts from damage.

5. Check that the shaft and housing are free from dust, dirt, metal burrs and other surface damage.

6. A light smear of oil or grease will help in assembling. If the bearing is likely to rust onto the shaft or housing, a light coating of 'Kopr-Kote grease' will make dismantling much easier, even in a year's time.

7. Locate the bearing square on the shaft or housing and with the pipe located on the ring or cup gently drive the bearing into place. Make sure the bearing does not cock askew and drive the bearing firmly home. Care must be taken not to use too much pressure as damage to the bearing or housing could result.

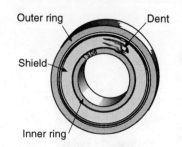

Figure 4.17 *Badly dented bearing shield or shell. (Courtesy New Departure) square*

> **Note: Be careful not to damage seals when driving bearings onto shafts or into housings. It will cause premature failure of the bearing**

Soft metal drifts (punches) can be used to drive off a bearing that is not going to be used again, but it is not advisable to use a soft metal drift when assembling a new bearing. There is a tendency for the soft metal to chip off and enter the bearing causing premature failure.

Seals

There are a range of different types of seals, all designed to either keep something in or out. In agricultural machinery they can be divided into:

- Static, where there is no movement between surfaces. These include gaskets, O-rings and sealants.

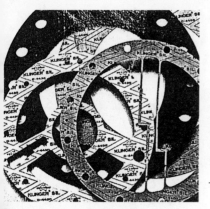

Figure 4.18 *Various synthetic fibre gaskets. (Courtsey Klinger)*

- Dynamic, which seal between moving parts. These include lip seals, packing, and mechanical end face.

Gaskets

Gasket material should be soft and pliable, enabling the gasket to follow surface irregularities. It should have a high degree of elasticity and flexibility to allow it to change shape as the bolts are tightened.

Gaskets are made from a range of different material, including treated paper, cork, rubber, compressed synthetic fibre, plastic, copper and various other soft metals.

Gaskets are usually used to make a gas-, steam-, or liquid-tight seal between parts of a machine, engine, or component. This could be sump onto engine block, water pump onto engine block or fuel gauge unit into fuel tank.

If any of these parts require dismantling a new gasket should be fitted when reassembling to make a good seal. If a new gasket is not available, care must be taken not to damage the old gasket. It can be re-used provided it is not damaged too much and a good quality sealant compound coating is given to both sides of the old gasket. These come in the form of a paste or 'goo'. Care must be taken to use the correct compound. Some will not stand heat, others will not stand pressure, or may damage synthetic gaskets. "Aviation form-a-gasket" is a good all round sealant and can be used in many situations. "Silastic" is another flexible type gasket sealant that should always be in a farm toolbox. Always follow the manufacturer's instructions on the container.

If a small gasket is damaged and cannot be used again, a new one can be cut from a piece of suitable gasket material. Even a cornflakes box can be used to make a gasket at a pinch. It should be thinly coated with gasket sealant before assembly.

To cut a gasket, first clean the old gasket from the face of the part. Lay the fresh gasket material on the face and hold it in position. With a small ball peen hammer gently tap around the outside edge of the part, so that the waste gasket material will fall away (See Figure 4.19). Locate any holes that need to be cut and repeat the previous operation until you have a gasket that will seal the parts when assembled. Make sure all holes and other parts of the gasket match the face of the part and clean any waste gasket material from the holes.

Figure 4.19 *Cutting a water pump gasket*

To install a gasket:

* Make sure both joint surfaces are clean.
* Bolts or studs should be clean, and the threads lightly lubricated.
* Insert enough bolts or studs in one face to locate the gasket. Make sure it lines up correctly and evenly.
* With the gasket in place bring the two surfaces together. Hold in place with bolts or studs, finger tight only.
* Moving around diagonally opposite studs tighten just firm the first time around, increasing tension each time around for three times. Remember to tighten diagonally each time.

O-rings

O-rings are usually made of rubber, neoprene or plastic. They are used in many situations where a static seal is required. Thay are used especially where high pressure is involved such as in hydraulic components, in hydraulic cylinders, control valves and hose couplings.

To be effective an O-ring must be elastic. It must be able to be stretched repeatedly to twice its normal length and still return to its original size.

Figure 4.20 *O-ring seal as used in a Rega low volume pump; the O-ring groove can be seen in the top cover*

Faults that can occur with O-rings include:

- Loss of elasticity due to overheating. Cracks will develop around the ring and the seal will leak.
- O-ring spongy, swollen or distorted. This could be due to exposure to chemicals or fuel when the O-ring is not suited to that chemical.
- Nicks, cuts or twisting. Probably due to incorrect assembly.
- Edges of the ring frayed, due to poorly fitting the ring or over-pressurisation in groove.

Make sure the correct size O-ring is used in relation to the size of the groove, that it sits correctly in the groove and it cannot get pinched between the two faces as the parts are assembled. Make sure the grooves and recesses are clean and free from burrs and sharp edges. The two faces of the part should be pulled down and tightened evenly.

Sealants and gasket compounds

Sealants are specially formulated to contain liquids or gases under pressure and can be used as an alternative to a gasket or O-ring in some situations where operating conditions are not too severe.

Sealants are available in various forms, to be used in different situations. It is always advisable to have some on hand to use in emergencies. There are three basic types of sealants classified according to their hardening properties:

1. Non-hardening, a soft mastic type material used when joining parts where little or no pressure is exerted on the joint.

2. Hardening-flexible. The majority of sealants fall into this category. They contain curing and setting adhesives but remain flexible when cured. Acrylic, silicone and neoprene are some examples. They are used where some flexing of a joint may take place and a rigid hardening sealant would crack.

3. Hardening-rigid. Epoxy, polyester and other resin compounds provide a seal that is rigid when cured.

Dynamic seals

All seals that prevent leakage of fluids, or entry of dirt and other matter, between rotating or sliding shafts and their housings are classed as dynamic seals. These include lip seals in gear boxes and sealed bearings, where only low pressures (below 5 psi) are present. Packing and mechanical end face seals are used where higher pressure is present, such as in hydraulic pumps or water pumps.

Lip seals

Radial lip seals are the most common seals found on farm machinery. They are used on gear boxes, wheel bearings, sealed bearings on disc and offset disc ploughs.

Most lip seals are made up of three parts; a metal case, the sealing element and lip, and the small spiral spring called a garter.

The sealing element or lip is usually made of synthetic rubber or leather. This is bonded to the metal case, and the garter spring forces the seal onto the shaft to make a seal.

The sealing ring is formed with a knife edge at the contact point which engages with the rotating shaft. A garter spring

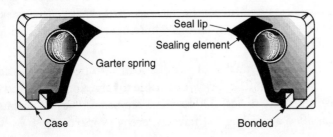

Figure 4.21 *Cross section of a typical lip seal, showing parts and construction*

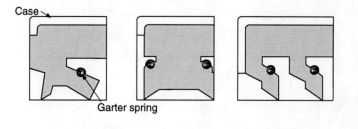

Case

Garter spring

Figure 4.22 *Various types of lip seals*

may be fitted inside this ring to help load (keep pressure on) the seal. Various materials are used for the sealing ring, depending on the operating conditions. The most common is nitrile which is compatible with most lubricants, or leather in applications where dust and dirt are a problem.

One of the most common causes of lip seal failure is damage that occurs when fitting the seal. Care must be taken and by following a few simple steps this damage can be avoided.

Fitting seals:

1. Check the seal size with shaft and housing.

2. Check the seal lip for damage. Any nicks, cuts or distortion will cause the seal to leak.

3. Check the shaft surface; keyways, holes through shaft, nicks, burrs or scratches. These can damage the seal when fitting. Especially check where the seal will be located, it must be very smooth at this point.

4. Make sure the seal faces the right way, towards the oil in a gear box.

5. If the seal is metal cased, a coating of gasket sealant will prevent leakage around the housing.

6. Use a properly designed tool to push the seal home in the housing. Push evenly on the outside ring only. Never press on the sealing ring or lip. Make sure the seal is not askew in the housing and it is pushed home firmly.

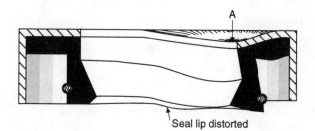

A

Figure 4.23 *Seal damaged in fitting. Punch damaged case at A, distorting seal face and ring*

Seal lip distorted

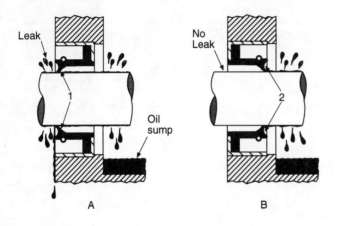

Figure 4.24 *Lip seals must face the fluid. Wrong (A) and right (B) way to fit seal*

7. If the seal has to be fitted over a spline, keyway or other sharp surface, a special protective cone or sleeve (mounting bullet) should be used. If these are not available use a piece of shim brass as a sleeve between the seal and the shaft.

8. Apply a coating of lubricant to the lip and the shaft so the seal slides freely along the shaft.

9. Turn the shaft by hand to make sure it runs freely.

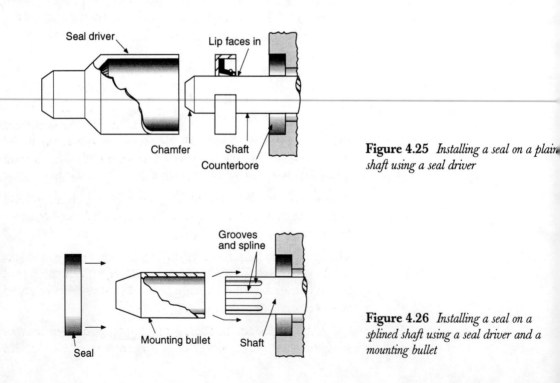

Figure 4.25 *Installing a seal on a plain shaft using a seal driver*

Figure 4.26 *Installing a seal on a splined shaft using a seal driver and a mounting bullet*

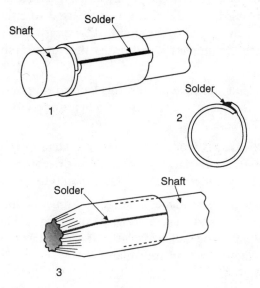

Figure 4.27 *Making a seal mounting bullet or sleeve from shim brass*

10. The seal may leak slightly until the lip wears in.

Most seal failure occurs from mechanical damage. If the seal has been properly fitted, the most likely mechanical damage is from the outside, in the form of material being allowed to build up around the seal. Things like baler twine, wire, weeds or straw can wrap around the shaft. As they wind tighter they will damage the seal enough to make it inoperative.

Packing seals

Packing seals are one of the oldest types of seals used. They consist of rings of soft yarn, like hemp, flax or cotton, lubricated with a dry lubricant such as graphite or molybdenum disulphide, that are fitted into a stuffing box and compressed by a gland nut or clamp plate. The sealing face is the full length of the packing (see Figure 4.29).

This type of seal is still used on some small water pumps and many irrigation pumps, taps, steam valves and some hydraulic fittings.

On water pumps some seepage (one or two drops per second) from the gland nut is necessary to keep the packing cool and lubricated. If the seepage becomes excessive the gland nut or clamp plate should be tightened.

On some of the larger pumps this lubrication is sometimes assisted with what is termed a "lantern ring", or "gland pressure feed fitting"; a special ring included when the packing rings are being fitted. This ring is connected to either the suction or discharge line of the pump, depending on the design of the system. It supplies liquid under pressure to lubricate the packing.

When pumping abrasive liquids, this lantern ring may have to be fed with clean water from another source to reduce wear in the gland.

If the gland leakage cannot be stopped by tightening the gland, it may be because the packing material has dried out or disintegrated. If this is the case the packing should be renewed.

Packing material is available in lengths, it must be moulded around the shaft and cut into rings of the correct diameter. Special graphite tape is also available. It can be used to form a packing ring to suit any sized gland.

To fit new packing:

- Make sure the gland is clean and all old packing is removed.
- Check the shaft for wear or scoring in the gland.
- Make sure the correct size and type of packing is used.
- If using continuous packing, cut packing into rings.
- Stagger the joints at least 90° for each ring used.
- Install the gland ring or follower, and pull up finger tight with gland nut or clamp plate.
- Start up the pump and allow the gland to leak to ensure it is lubricated. Then tighten the gland until leakage is reduced to one or two drops per second.

Figure 4.28 *Packing seal material can consist of various types of material, including plaited hemp impregnated with a graphite grease and compressed into a square section, alternate layers of reinforce rubber and duck (canvas), or tallow soaked flax packing ("greasy hemp")*

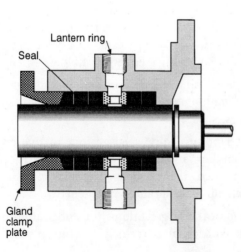

Figure 4.29 *Packed gland from a Southern Cross irrigation pump*

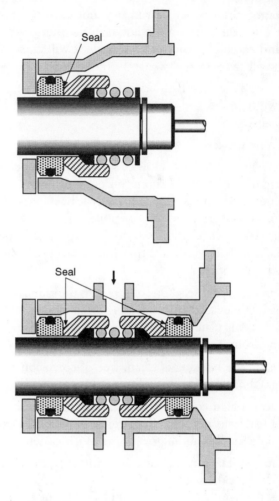

Figure 4.30 *Mechanical end face seals from Southern Cross irrigation pumps. Top, single mechanical seal and bottom, a double mechanical seal, with a lantern ring for seal lubrication*

Mechanical end face seal

Most small water pumps, domestic pressure systems, bushfire pumps and some larger irrigation pumps are now fitted with mechanical end face seals. The sealing faces are made up of different materials, with various combinations available. Some of the common ones are stainless steel/carbon, lead bronze/ carbon, or ceramic/carbon.

The carbon used for the seal face is usually a composition of carbon and a metal or resin filler. It is very brittle and if distorted in any way will crack and leave the seal ineffective.

If a mechanical seal leaks, it cannot be tightened or adjusted in any way. It will need to be replaced with a new one. The whole seal should be replaced, it is not advisable to only replace one side of the seal. They are usually supplied in matched pairs.

The pump must never be run dry, not even for a few seconds. The seal relies on the liquid in the pump for lubrication and cooling. If the seal overheats it will craze or crack which will cause it to leak.

Some sheep dips, jetting fluids and wettable powders contain abrasive material and these can cause rapid wear of a mechanical seal.

To fit a new mechanical seal:

1. Dismantle the pump. How this is done will depend on the design of the pump. In some pumps it can be done by first removing the pump housing from the backing plate. Then remove the impeller, (a gear puller may be required for this task). The old seal should now be visible around the shaft, the half with the spring and rubber may come away with the impeller.

2. Remove the old seal. On some pumps the old seal is removable without further dismantling. Most pumps will require the backing plate to be removed from the shaft and the seal driven out, using a punch through the shaft hole.

3. Ensure the housing is clean, free of corrosion and burring. Check the O-ring (if fitted) came out with the seal.

4. Smear a light coating of lubricant in the housing and on the seal housing. Kopr-Kote grease will prevent corrosion and make dismantling next time much easier.

5. Install seal by pushing on outer ring only, taking care not to damage the sealing surface in any way. Make sure the seal does not cock askew and push it home firmly.

6. Reassemble the backing plate, making sure it is centred on the shaft.

7. Carefully place the other half of the seal on the shaft and hold in place with the impeller. Make sure the spring and rubber are free and not pinched anywhere.

8. Make sure the new seal turns freely with the impeller.

9. Replace pump housing; clean off any old gasket material or sealant. If an O-ring seal is used, inspect it for damage, twisting or distortion. Replace if necessary. If re-using an old gasket, it is advisable to use some gasket sealant. Replace pump housing and tighten studs or nuts finger tight, then tighten alternately a little at a time until tight.

10. Fill pump with water before operating to avoid seal damage.

CHAPTER 5

Power transmission devices

Power is transmitted in many ways from the power source to the driven machine. In this chapter we will examine just a few, and investigate some problems that can arise with these devices.

Flat belt and pulleys

Flat belts and pulleys are not in wide use today, but shearing sheds are one exception. Any shearing shed using shaft driven overhead gear will probably have a flat belt drive from the engine, or electric motor to the shafting.

When servicing a few fundamentals should be remembered:

- Pulleys should always line up, with shafts parallel to each other.
- Flat pulleys should have a crown (convex shape) on the driving face. The belt will run up to the middle of the crown, thus running in the middle of the pulley.
- Bearings on shafting should be checked (see Chapter 4, Bearings and seals).

Belts can be made of leather, rubber or a synthetic material. Depending on the material there are several ways of joining flat belts. The two most common are:

- Lacing, using either leather or rawhide
- Various types of metal joining devices.

V-belts and pulleys

V-belts and pulleys are a very popular way to transmit power because they are clean and require no lubrication. They are smooth starting and running, and they cover a wide horse

Figure 5.1 *Examples of flat belt pulleys*

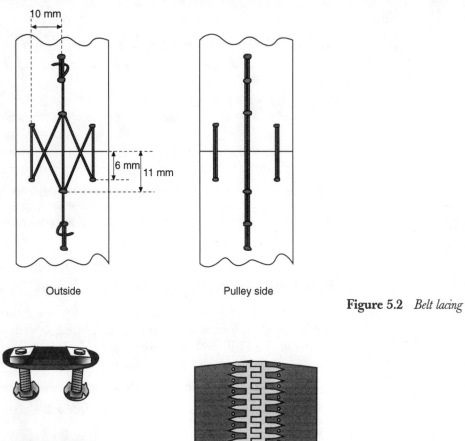

Outside   Pulley side

Figure 5.2 *Belt lacing*

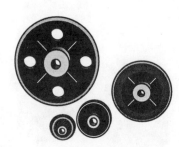

Figure 5.3 *Various flat belt joiners*

power range. They tend to dampen vibration between driving and driven machines, and are therefore quiet. They act as a slip clutch in the power train as they refuse to transmit a severe overload of power, except for a very brief time. If a few simple rules are followed they are also usually trouble free.

V-belt pulleys are also sometimes referred to as "sheaves". They come in a wide range of diameters, with single or multi-Vs. They also have different sized Vs or grooves. They can be made of aluminium alloy, cast or pressed steel. The size of the grooves or "section" is denoted by the letters A, B and C, these three sizes being the most common on farm machinery,

Figure 5.4 *Selection of alloy V-belt pulleys*

Figure 5.5 *Step pulleys*

and A section being the smallest. Some special purpose belts are made with teeth, as in timing belt pulleys. Other pulleys can be much larger than C section; examples are the drum drive belt on some headers or some baler drive belts.

The single belt drive is the most common type of V-belt drive. It has a single groove in the pulleys, and a single V-belt used to transmit the power.

If regular changes of speed are required, matched step pulleys allow this with little time loss. A quick tension adjustment is usually provided, so it is only a matter of slackening the belt tension, changing to another drive ratio, and retightening the belt tension.

In order to transmit more power through a V-belt drive it is sometimes necessary to increase the number of belts used. These arrangememnts are classed as multi V-belts.

Banded V-belt drives overcome the tendency of belts to whip, twist or jump off the pulleys. On a banded V-belt the V sections are vulcanised (hardened) onto a common band.

Adjustable link V-belts

Adjustable link V-belts are available in A and B section, in lengths up to 50 metres.

The advantages of this type of belt are:

Figure 5.6 *Adjustable link V-belt on a multi V-belt drive*

- It eliminates the necessity of having to keep a number of spare belts of varying lengths.
- They have a higher power transmission rating than a standard V-belt of the same cross section code.
- No adjustment of shafting or pulleys is required as belts wear or stretch. Just take a link out.
- They are more flexible and so can be used on smaller diameter pulleys without belt damage.

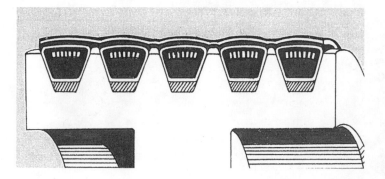

Figure 5.7 *Banded V-belt*

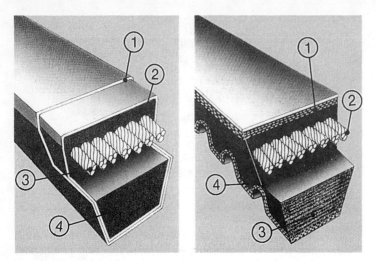

1. Cover. Woven cotton fabric impregnated with neoprene.
2. Tension section. Synthetic rubber especially compounded to stretch as the belt bends around a pulley.
3. Cords. High-strength, synthetic fibre cords carry the horse power load and minimise stretching.
4. Compression section. Synthetic rubber compound to support cords evenly and compress while bending around pulleys.

Figure 5.8 *V-belt construction* – left, *a Wrapped belt and* right *a Raw Edge*® *Cog-Belt*®. *(Courtesy Dayco)*

- They can be run on multi-belt drives, avoiding the necessity of replacing a matched set if one belt breaks. Just replace the broken link.

The disadvantage of adjustable link V-belts is that they are very expensive. One has to weigh up the advantages against the expense when replacing V-belts.

Maintenance of V-belts and pulleys
- Keep oil and grease away from rubber belts.
- Before doing any maintenance work on any power drive make sure the engine is turned off, and remove the key to avoid accidental starting. Never attempt to adjust a belt while the machine is in operation.

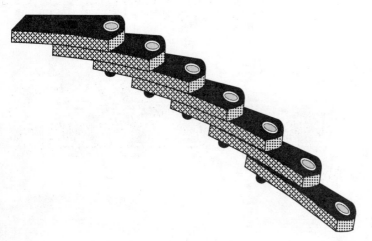

Figure 5.9 *Adjustable link V-belt*

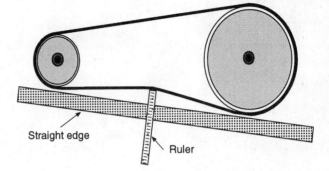

Figure 5.10 *Checking V-belt adjustment*

- Maintain correct V-belt tension. A loose belt will slip, causing the belt to burn in one particular place, (see Figure 5.11). A tight belt will overheat, stretch and damage bearings, pulleys, and shafts.
- The operator's manual should give you the correct tension for the belt on the machine you are servicing.
- Most agricultural V-belts should have from 10 to 25 mm of slack from the straight belt when pressure is applied, depending on the length, belt section and horse power transmitted.
- If the pulley is worn it will need to be replaced, and the belt re-tensioned.

Insufficient belt tension vies closely with worn pulley grooves as the leading cause of V-belt slippage and other problems. If unsure of correct tension use the sight, sound and touch method.

Sight: While the drive is operating look for a slight bow or sag in the slack side of the belt. This is normal and should appear more noticeable under load such as start up, or load cycle as in the plunger of a bailer.

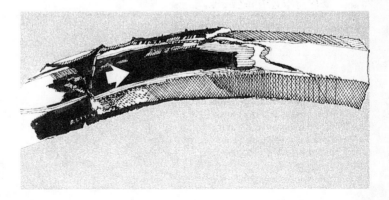

Figure 5.11 *Belt burn caused by loose belt. (Courtesy Dayco)*

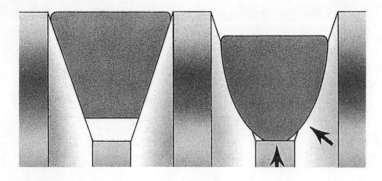

Figure 5.12 *A V-belt should never be run on the bottom of the pulley groove. This is caused by worn belt and excessive dishing or wear of pulley. Left is a correct fitting belt and* right, *a belt running in worn (dished) sidewall grooves. (Courtesy Dayco)*

Sound: Properly designed and adjusted V-belt drives should not squeal or howl under peak load conditions. If necessary, stop and start the drive again. If a squeal is heard, the belts should be tightened just to the point where they do not squeal under peak load. Newly installed belts require about 24 hours to become fully seated in the pulley grooves, so a little extra tension when installing, and a recheck of tension the next day is good practice.

Touch: V-belts do not always squeal when slipping. If slipping is suspected, a sure way of determining it is to turn off the engine and place your finger on the inside of the pulley groove. A slipping belt will generate enough heat so that it is too hot to keep your finger there, provided there is no outside heat source. Assuming the pulley grooves are not worn as in Figure 5.12 the belts should be tightened.

Faults, problems, causes of belt breakage

Seasonal machines that are left outside between seasons should have all perishable items on the machine, such as V-belts, removed. These should be stored in a dark dry storage area. If they are left on the machine they will deteriorate rapidly, as sunlight, heat and cold adversely affects these components. Deterioration of the rubber and cords could cause the belts to break when working in the next season.

Never fit a belt onto a pulley using a lever or screwdriver. This can and usually does break some of the load-carrying tensile cords in the belt. When this happens, the belt may either break or turn over in the groove, usually in the first few minutes of operation. Always loosen the adjustment to assist with removal or replacement of belts.

Misplaced slack can also cause belt breakage, usually on start up. This occurs on multi-belt drives when all the belt slack is

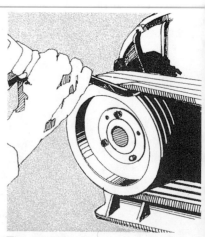

Figure 5.13 *Never use a screwdriver or lever to install a V-belt. (Courtesy Dayco)*

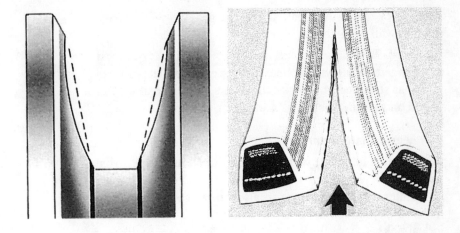

Figure 5.14 *Worn groove sidewalls and cut tie-band caused by worn grooves. (Courtesy Dayco)*

not brought to the same side of the drive before tensioning. If some belts are tight on one side, and others are tight on the alternate side, the heavy shock load of starting will be taken by only some of the belts. This can weaken or break the load carrying cords.

Inspection of pulleys and grooves and alignment are very important to the life of a V-belt. Particular attention should be given to the following conditions:

Worn groove sidewalls. This can cause belt slip. If extra tension is used it can cause belts and bearings to overheat, and premature failure of both. V-pulley templates are used to accurately check the wear in grooves. Hold a torch behind the template and the amount of wear will easily be seen. Dishing should not exceed 1 mm for individual V-belts, or 0.5 mm for banded belts. Worn grooves cause banded belts to ride too low in the groove and the result is cut tie-bands.

Shiny groove bottom. This indicates a worn belt or pulley, or both, with the belt running on the bottom of groove instead of having a wedge grip on the side. (See Figure 5.12).

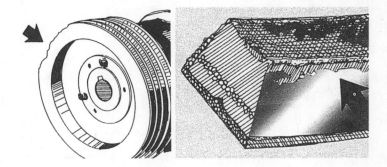

Figure 5.15 *Damaged grooves and sidewall damage caused by damaged grooves. (Courtesy Dayco)*

Damaged grooves. Any damage on the grooves will damage the side wall of the belts.

If a *multi-belt drive* is being used and one belt becomes stretched or damaged, replace all the belts. If only one is replaced, it will not be as stretched or worn as the others, and the extra strain placed on that single belt will damage or break it.

Pulley and shaft alignment

Pulleys must be in line with each other in all directions. Misaligned pulleys can cause rapid wear of the V-belt sidewalls and pulleys. Misalignment can also cause separation of the tie-band on banded belts, or apparent mismatching of individual belts.

The three basic types of pulley misalignment are shown in Figure 5.16.

Alignment should be checked and corrected in the order horizontal/angular, vertical/angular, then parallel.

Horizontal/angular: shafts in the same horizontal plane, but not parallel.

To check, use a straight edge or stringline across the face of the pulley near the centre.

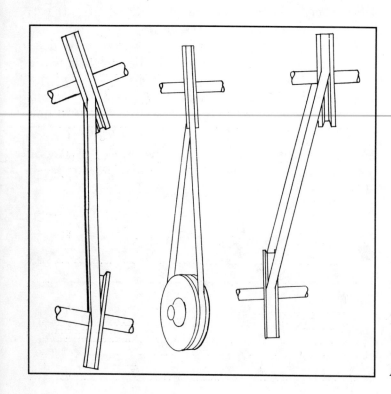

Figure 5.16 *Three types of pulley misalignment* – left, *horizontal angular,* centre, *vertical angular and* right, *parallel. (Courtesy Dayco)*

To correct, loosen motor or bearing mounting bolts and move shaft until all four points are touching the straight edge.

Vertical/angular: shafts are parallel but pulleys are not in line.

To check, place straight edge about 25% of the radius from the outside diameter of both pulleys.

To correct, use shims under one bearing housing or one side of motor base depending on correction required.

Parallel alignment: shafts are parallel, but pulleys not in line.

To check, use a straight edge or string near the centre of the pulley.

To correct, loosen locking device on the pulley, and slide the pulley along the shaft until all four points meet.

Note: Pulleys should be mounted as close as possible to the bearings, to reduce overhung load on bearing and "shaft whip". Relocation of both pulleys or bearings may be necessary.

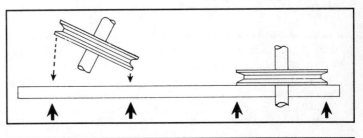

Figure 5.17 *Checking horizontal/angular alignment. (Courtesy Dayco)*

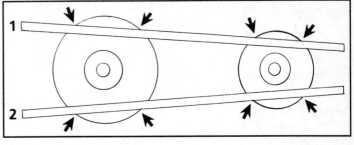

Figure 5.18 *Checking vertical/angular alignment.*
1. Repeat on opposite side of shaft.
2. Straight edge should touch the points arrowed. (Courtesy Dayco)

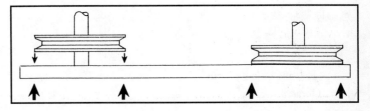

Figure 5.19 *Checking parallel alignment. (Courtesy Dayco)*

Table 5.1 *V-belt drive troubleshooting guide. (Courtesy Dayco)*

Symptom \ Cause	Belt levered on or misplaced slack	Belt rubbing on guard	Pulleys misaligned	Worn or damaged grooves	Pulley too far from bearings	Poor bearing or shaft condition	Insufficient tension	Excessive tension	Improper pulley installation	Belts worn in normal service	Wrong belt cross section	Mismatched belts or mixed brands	Machine induced impulse or shock	Improper or prolonged storage	Excessive heat	Excessive oil or grease	Use of belt dressing	Abrasive environment	Foreign objects in grooves	Excessive moisture	Overloaded drive underbelting	Drive seriously overbelted	Pulleys too small	Insufficient wrap on pulleys	Backside idler
Rapid sidewall wear		O	O	×			O							O	O	O	O	O	O						
Worn cover on back		×																							O
Belt turns over or runs off	O						O				O		×						O						
Belt soft, swollen																×	O								
Belt slips, squeals (spin burn)				×			×				O					O					O	O		O	
Belt cover split	×																		O						
Underside cracked				O										O	×								×		×
Tie-band damaged		O	O	×															×						
Repeated breakage	O							O					O						O		×				
Belt rides too high										×						O									
Belts bottoming				×						O	O														
Repeated tightening necessary				O			O				O												×		
Belts vibrate or appear mismatched			O	O			O	O				O	×									O			
Bearings overheating				O	O	O		×								O						O	O		
Shafts whip or bend				O	O	O		×														O	O		
Cracked bushing				O					×																
Pulleys wobble				O		O			×																

× Indicates most common causes

O Indicates other possible causes

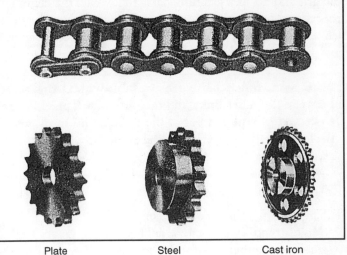

| Plate | Steel | Cast iron |

Figure 5.20 *Standard roller chain and various sprockets*

Chains and sprockets

Chains and sprockets are a more positive drive than V-belts. They maintain a constant speed ratio between the driving and driven sprockets. They also require more maintenance.

They are used where different components have to be correctly timed, such as the needle of a baler being timed in relation to the movement of the plunger.

Two main types of chain are used and they have to have matching sprockets:

1. Roller chain
2. Detachable link chain also known as agricultural or hook link chain

Roller chain

Standard roller chain is made up of alternate roller links and pin links. The rollers are free to rotate on the bushing pressed into each roller link. This reduces the rubbing action between the chain and the sprocket as the chain link rolls onto the sprocket, thus avoiding wear from sliding friction. Because each roller and bushing functions like a plain bearing, lubrication is essential to the operation of chain drives.

However in some situations, particularl. (Courtesy Dayco)y in gritty and dusty conditions such as potato diggers working in sandy or abrasive soil, lubrication only tends to hold the grit and act as a grinding paste that wears the chain link rollers and bushing much quicker. It is sometimes better to leave the chains dry, and put up with the noise rather than just give them a splash of oil once a day.

> Note: Some roller chains are available with O-ring seals between the roller links and the pin links. The seals are to keep dust and dirt out of the rollers, and in so doing prolong the life of the chain. The extra cost of this type of chain is not always recouped in longer chain life.

Roller chain is available in:

1. Single strand, simple or simplex
2. Multi-strand. Two strand or duplex, three strand or triplex; etc. The number of strands required will depend on the power to be transmitted.
3. Double pitch roller chain
4. Agricultural roller chain.

Standard roller is widely used on agricultural machinery.

Multi-strand chain is used in some applications to transmit more power without increasing the pitch or speed of the chain.

Double pitch roller chain is used to transmit light loads at slow speeds. The pitch is twice that of a standard chain. It will run on a standard roller chain sprocket of the same size, or special sprockets are made for double pitch chain. Double pitch chain is cheaper than a standard roller chain of the same size.

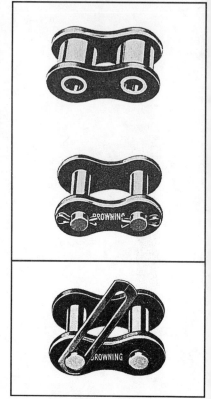

Figure 5.21 *Roller chain parts* – top, *roller link;* centre, *cotter pin connecting link and* bottom, *spring clip connecting link*

Roller chain is classified by:

- Pitch, which is the distance from the centre of one pin to the centre of the next pin.
- Roller diameter, the diameter of a new roller.
- Roller width, the width of a roller between the two side plates.
- Pin diameter should also be checked if only buying a connecting link.

All these dimensions must be stated when ordering a new chain.

When designing chain drives it is normal practice to use a chain with an even number of links and a sprocket with an odd number of teeth, or vice versa. This will ensure that the same link does not contact the same tooth each revolution, thus helping to reduce uneven wear.

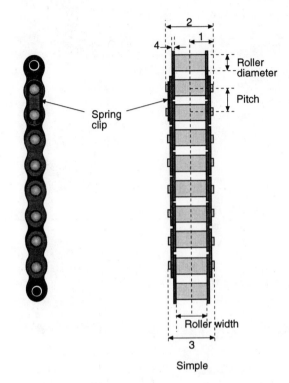
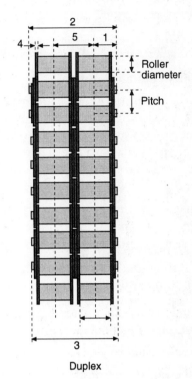

Simple

Duplex

Figure 5.22 *Simple and duplex chains showing the dimensions required when ordering. These dimensions are critical in multi-strand chain*

1. Distance outside of chain to centre of first roller
2. Length of pin, joiner link
3. Length of pin, standard link
4. Plate thickness
5. Distance centre to centre of next roller in multi-strand chain only

Roller chain maintenance

- Alignment and correct tension are critical to the proper operation of the drive. See Figures 5.17, 5.18 and 5.19 for the correct alignment proceedures.
- Chain drives should be protected from dirt, where possible and practical.
- Chains should be properly and adequately lubricated. An oil can or brush is the most simple way or lubrication can be supplied with a drip lubrication system, automatic spray or oil bath for high speed drives. Whatever lubrication system is used, the lubricant must penetrate into the roller and bush. The higher the chain speed the greater the supply of lubricant required.

- New chains should not be installed on badly worn sprockets. It will only increase the wear on the new chain. Sprockets may be reversed on the shaft where practical, this may extend their life if it is done before they are too worn.
- Chains should always have proper guards fitted.

Roller chain adjustment

Roller chains are joined with a special connecting link. If the chain has to be removed this link has to be disconnected. The easiest way is to use a pair of pliers in the manner shown in Figure 5.23. If using a screwdriver for this job, care must be taken that the spring clip does not flip into the distant yonder.

Chain drives must be properly tensioned. If a chain is too tight it will bind on the sprocket and wear rapidly. If it is too loose it will tend to whip, which also causes wear and vibration.

For chains with adjustment on the shafting, the slack should be adjusted to approximately 2% of the shaft centre distance. For instance, if the shaft centre distance is 1 m the slack or sag in the chain should be 20 mm, measured with a ruler and straight edge.

If the adjustment on the shaft is limited and the correct tension cannot be attained, a link may have to be removed. This will require the removal of two links, a roller link and a pin link, otherwise reconnecting the chain will be impossible.

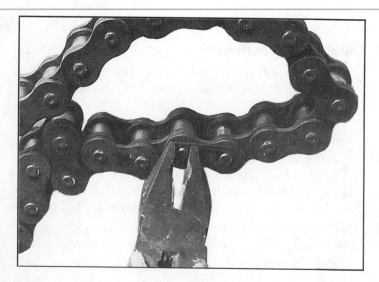

Figure 5.23 *Removing spring clip from connecting link*

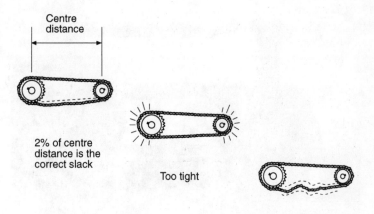

Centre distance

2% of centre distance is the correct slack

Too tight

Too loose

Figure 5.24 *Correct chain tension*

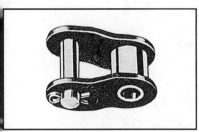

Figure 5.25 *Special offset or half link*

If the removal of the two links leaves the chain too short with the limited adjustment, special half links are available to overcome this problem (Figure 5.25). They are connected to the end of the chain and the chain is then joined with the usual connecting link.

For drives with fixed shafts, an adjustable idler sprocket will be used to adjust the slack. Sometimes this idler sprocket will be spring loaded and it will automatically take up the slack.

It is important to remember that chains need to be readjusted from time to time. Chains do not stretch; the lengthening of the chain is due to the wear in the rollers, bushing and pins, not to any physical deformation of the chain links.

A chain detaching tool will make link removal much easier.

Roller drive chain faults and problems

Chain jumping or climbing teeth on the sprocket. This could be due to worn chain, loose pins in side plates, chain too slack, or excessive build up of dirt or grease in sprocket teeth. A chain with a broken roller bush will tend to click and ride up on a sprocket tooth each time that link comes around. If the cause is a worn or a damaged chain, replace the chain before the sprockets are damaged .

Chain whipping, is usually from too much slack, or pulsating loads. It may also occur if some of the links in the chain have become stiff.

Tension sprocket

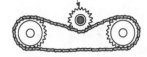

Figure 5.26 *Separate tensioning sprocket (idler)*

Chain breakage. It is rare for a chain to break completely, but it may occur in conjunction with other problems, like overloading the chain. A very worn chain may jump a tooth on

Figure 5.27 *Chain link detaching tool*

the sprocket which causes excessive tension and the side plates may fail.

Chain overheating is an indication that there is inadequate lubrication, the load is too great, the operating speed is too high, or all of the above. Chain drives are generally suitable for speeds of up to 1350 metres per minute.

Misalignment. Chain drives are more sensitive to misalignment than belt drives. Misaligned shafts or sprockets will result in rapid wear on the sides of the teeth, or wear on the chain side plates (see the section earlier in this chapter on the misalignment of V-belts).

Agricultural chain

Steel detachable link chains are used extensively on agricultural implements, both for transmitting power and for elevators. It is the least expensive type of chain and sprockets, because the sprockets are usually rough cast iron. The only machining necessary is the shaft hole and keyway.

This type of chain is well suited to moderate loads at slow speeds, not exceeding 120 to 250 metres per minute. In dirty and dusty conditions it is subject to wear, but because of the slow speeds involved wear is minimal. Detachable chains are not usually lubricated because the lubricant tends to hold the dirt in the joint.

Types
Agricultural chain is supplied in two types:

1. Detachable link chain, which is available in malleable cast iron or pressed steel. Both are detached in the same manner.

Figure 5.28 *Detachable link chain showing classification numbers*

The link to be detached is folded around and tapped with a hammer, driving it out of the hook of the other link. New pressed steel links are usually more difficult to detach than cast iron links. This type of chain is classified by number.

2. Agricultural roller chain, with pitch matching the detachable link chain and designed to run on the same sprockets. It is suitable for slow speeds, can transmit more power, runs quieter and smoother and does not wear as fast as the detachable link chain. It is not as durable as standard roller chain but much cheeper. It needs to be kept lubricated.

This type of chain is often made with lugs, suitable to take elevator brackets.

Maintenance

Little maintenance is required, provided that the chain is correctly tensioned. Because the chains are only slow moving, tension is not as critical as with roller chain, except if the machine is working on the side of a hill when a loose chain may run off the sprocket. If this happens there is every chance the chain will catch, wind around the shaft and break.

To correctly put a detachable chain on a sprocket, the split in the hook must be to the outside and the driving sprocket must transmit the power to the back of the hook (Figure 5.30).

Figure 5.29 *Detaching a hook link agricultural steel chain.* Top, *makeshift chain holder in vice;* Centre, *one link folded and driven out;* Bottom, *reconnecting chain*

Figure 5.30 *The correct fitting of a detachable link chain. The sprocket should rotate in a clockwise direction*

Gears, open and closed

All gear drives no matter what type will give a constant speed ratio between the driving gear and the driven gear. Most gears run on a shaft enclosed in a gear box containing lubricant, although some slow moving gears are classed as open gears and are not enclosed in a gear box.

There are many types of gears (see Figure 5.31), classified by:

- shape or cut of the teeth
- angle and position of the shafts.

Spur gears

Spur gears are the simplest form of gear. The teeth are cut straight across the face and they are always parallel to the shaft. Spur gears can be either closed (enclosed in a gear box and run in oil), or open (operated in the open without the protection of a gear box or oil bath lubrication). Many slow revolving gears on farm machinery are open spur gears. Some need to be lubricated and oil or grease can be applied at varying intervals. Some oil companies make a special open gear lubricant. If working in very abrasive dusty conditions it is sometimes better not to lubricate them at all, as the oil or grease only holds the dust and it tends to turn into a grinding paste.

Helical gears

The helical gear is so named because the teeth are cut at an angle to the face. The geometric figure involved is called a helix.

The helical gear runs quieter than a spur gear because there is constant engagement of the teeth and no sudden transference of load from one tooth to the next as in spur gears.

Most helical gears are enclosed and use oil bath lubrication.

Bevel gears

Some forms of bevel gear are used when the power has to turn a corner. The simplest form of bevel gear is the straight cut bevel gear, and as with the straight spur gear, it may be of the open or closed type. The small gear is usually called the "pinion" and the larger called the "crown wheel". All the same conditions as the spur gear apply to the bevel gear as far as lubrication and dust are concerned.

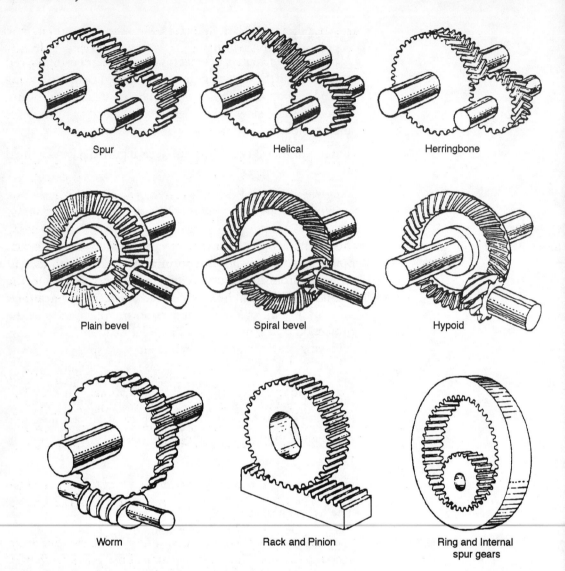

Spur Helical Herringbone

Plain bevel Spiral bevel Hypoid

Worm Rack and Pinion Ring and Internal spur gears

Figure 5.31 *Some gear types (from Hunt & Garver)*

Spiral bevel gears

These perform the same task as the spur bevel gear, but are quieter and smoother. They are very similar to the helical gear except the teeth. Beside being at a angle to the face, they are cut in a curve or a spiral. Spiral bevel gears are usually enclosed with oil bath lubrication.

Hypoid gears

At first glance they resemble a spiral bevel gear, but they differ considerably as the axes do not intersect. The axis of the pinion is offset to the crown wheel, so the teeth cannot be cut in a true spiral and still mesh properly. The teeth are cut in a complicated geometric figure known as a "hyperbolic paraboloid" from which the name "hypoid" comes. Most differentials use hypoid gears and require special hypoid oil in the housing.

Worm gears

Worm gears provide a wide range of ratios for reduction drives. They can be as high as 70:1. The crown wheel, (or sometimes ring gear teeth) is convex to provide greater contact area and greater strength. The worm teeth are continuous and resemble a screw thread. They impart a sliding action, and in effect it is the same as screwing a bolt into a thread, except that the bolt is stationary and the gear must turn as the teeth of the worm revolve. Some worm gears are open but most will be enclosed in a box with oil bath lubrication.

Maintenance

Lubrication is essential in all gear drives. With most gears being enclosed in a box it is a simple matter to keep the oil level at the manufacturer's recommendations. This level is usually only slightly above the bottom of the gears. As the gears rotate they carry the oil around lubricating the remaining gears and bearings. Always use the manufacturer's recommended grade of oil.

To prevent the oil escaping through the bearings there will be an oil seal. If the oil level is too high extra pressure will be placed on this seal and it may fail.

Any material (baler twine, weeds, straw, wire, etc) that may wrap around the shaft on the outside of the gear box should be removed. It will damage the seal allowing the lubricant to escape.

Open gears also require lubrication. It can be supplied by oil can, dripper, or spray. Special open gear lubricant is available from some oil companies.

Excessive wear in the bearings will result in "run out" and/or excessive backlash, (Figure 5.32) causing wear of gear teeth. It should be corrected as soon as possible.

As the gear teeth wear the small metal particles and other contaminants (dirt, water condensation) will collect in the lubricating oil. The oil should therefore be changed at regular intervals.

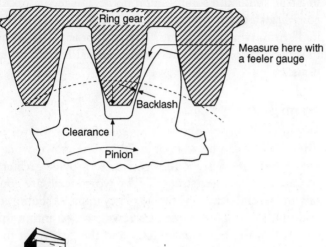

Figure 5.32 *Shows gear clearance backlash (measured with a feeler gauge)*

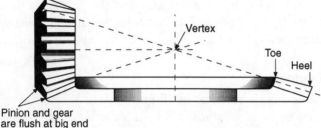

Figure 5.33 *Correctly adjusted bevel gear*

The clearance between gear teeth is known as gear "backlash". This clearance varies with gear types and applications, but when it becomes excessive, rapid gear teeth wear will result.

In many cases the backlash is not adjustable, in others it may be adjusted by moving shaft and bearing, or moving gears on the shaft. This can be the case in some bevel gear drives, where it is just a matter of loosening a set screw and sliding the gears closer together, or using spacers or shims behind the gears to hold them together.

When adjusting backlash, every effort should be made to maintain correct geometry in the meshing gears, particularly in bevel gears. Pitch lines of bevel gears should meet at the vertex when properly adjusted.

PTO and universals

The power take off (PTO) is a means of transmitting the tractor engine power to an implement connected to the tractor either by the drawbar or three point linkage.

Figure 5.34 *A power take off shaft with telescopic guard*

The PTO shaft of the tractor is usually located at the rear of the tractor, although some tractors are also equipped with front- or mid-mounted PTO shafts.

There are two main types of PTO tractor shafts:

1. The 6 spline, 35 mm diameter shaft, designed for either 540 rpm or 1000 rpm. The speed is selected by moving a lever near the operator's seat.

2. The 21 splined, 35 mm diameter shaft designed specifically for 1000 rpm. This type of shaft is usually fitted into the rear of the tractor and held in place with a circlip. In some tractors it is possible to change the 21 spline shaft with a 6 spline shaft, by simply removing the circlip, removing the 21 splined stub shaft and replacing it with a 6 splined stub shaft using the same hole.

Another PTO drive available on some tractors is the "ground speed", where the PTO is driven off the rear axle rather than the engine. It can be used where it is more efficient for ground speed rather than engine speed to regulate the output of the driven machine. It is also very handy in situations where the machine can be turned backwards, by reversing the tractor a short distance. This sometimes helps to clear blockages of a minor nature.

To connect the PTO shaft of an implement to the tractor, care must be taken that the yokes of the PTO are on the same plane (lined up correctly). On most PTO shafts this is made easy with the telescopic shaft being either rectangular or a square shaft with a projection and matching groove, so the yokes cannot be connected the wrong way.

Figure 5.35 *A reversible stub shaft which can provide either 540 or 1000 rpm (Case)*

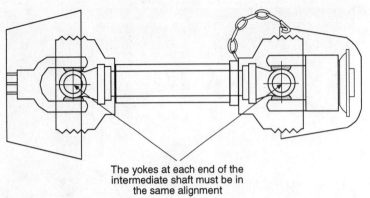

The yokes at each end of the intermediate shaft must be in the same alignment

Figure 5.36 *The correct yoke alignment on a PTO*

Universals

Universals are the means of bending the PTO shaft drive on a trailing implement to allow it to turn a corner, or a three point linkage implement to be raised and lowered. If the angle of the universal exceeds its recommended angle while rotating, damage can occur. The PTO should always be disengaged when turning sharp corners, or when raising three point linkage implements.

Maintenance of universals

Most agricultural universals are fitted with a grease nipple allowing regular lubrication. They should be given two or three pumps of multi-purpose grease daily. Too much grease will pop the dust cap allowing dust and dirt to enter the needle bearing in between the yolk and the cross. Dirt in this bearing will very quickly ruin the bearing and a new cross and bearing will need to be fitted.

Fitting a new cross and needle bearing to a universal

Tools required are hydraulic press or vice, or brass hammer and a vice, a piece of round steel 25 mm long (2 mm smaller in diameter than the bearing cup of the universal), a piece of pipe 12 mm long (2 mm larger than the bearing cup), large and small sockets (used in place of the previous two items provided clearance is maintained between sockets and cups and socket and yoke), circlip pliers, and vice grips (Figure 5.37).

> **Note: If the cross or bearings are to be used again, make sure the universal is greased well. Roll the universal around several times to spread the grease in the cups. The grease will hold the tiny needle roller bearings in the cup when it is removed, instead of loosing them all over the workshop floor.**

Figure 5.37 *The spacer,* left, *and the* round rod, right, *required to change the cross in a universal. An internal circlip is* on *the outside end of the cross*

Remove the locating circlips. They may be internal in the yoke on the outside end of each cross, or external on the inside of the yoke, with the circlip around the bearing cup (Figure 5.38).

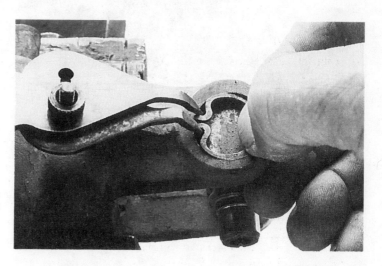

Figure 5.38 *Removing circlip with circlip* pliers. Thumb will stop circlip from slipping

Using a press or a vice and the 25 mm long round rod and pipe (or the large and small sockets), press one side cup out of the yoke, by pressing the whole cross and the bearing cups through the yoke until about half the bearing cup is through the yoke (Figure 5.39). This cup can be removed by grasping it with the vice grips and pulling it out with a slight twisting movement (Figure 5.40).

Alternately, gently grip the cup in the vice and hit the yoke with the brass hammer to drive the yoke away from the cup

Fig 5.39 *Using rod and pipe or sockets to push cups through yokes*

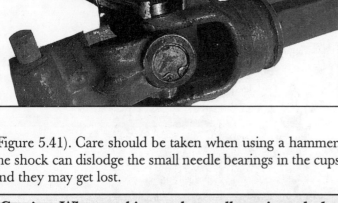

Fig 5.40 *Using vice grips to twist cup out of yoke*

(Figure 5.41). Care should be taken when using a hammer, the shock can dislodge the small needle bearings in the cups and they may get lost.

> **Caution: When working on the smaller universals do not push the cups through too far. It is difficult to start the second cup back in the yoke from the inside.**

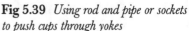

Fig 5.41 *Driving cup out of yoke with brass or soft faced hammer*

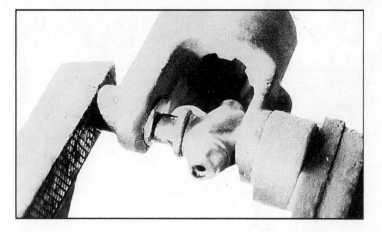

Figure 5.42 *Pushing the last remaining cup through yoke (using round steel rod and spacer)*

The second cup now has to be pushed back through the yoke. Turn the universal over and using the round steel on the exposed end of the cross and the pipe spacer on the outside of the yoke (to take the cup as it is pushed through), push the other cup out and remove it in the same manner (Figure 5.42).

The cross can now be removed.

> **Note: On some of the larger universals the cups can be pushed completely through. There is enough room to remove the second cup from the inside, and then remove the cross.**

Repeat this process for the other two legs of the cross.

Inspect the yokes for any damage and remove any burrs.

Take the new universal cross and carefully remove all the cups.

Make sure all the needle rollers are in place. They may need to be held in place with a smearing of heavy grease. (Do not use high temperature wheel bearing grease for this job, although it does a very good job holding the needle bearings in place, it may cause problems later if the grease separates when mixed with the usual grease used for greasing the universal. See Grease in Chapter 2.) If one needle roller becomes dislodged in assembly, (it usually ends up in the bottom of the cup and the circlips holding the cup won't fit in place), the whole thing has to be dismantled, the needle replaced and reassembled. Even worse the needle or cup can be damaged requiring a whole new universal cross.

Lightly lubricate the cups and the holes in the yoke.

Using the press or vice very carefully start one of the cups into one side of the yoke making sure that it is straight and

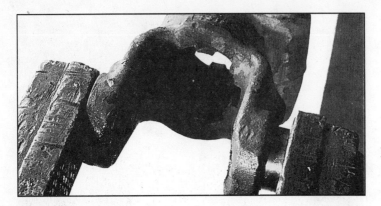

Figure 5.43 *Fitting the first cup into the yoke, do not push more than half way in at this stage*

square. Push the cup about half way into the yoke (Figure 5.43).

> **Caution: Extreme care must be taken when pressing cups into or out of the yokes. Too much pressure can cause distortion to the yokes or damage to the cups.**

Take the cross and place one of the legs through the empty hole in the yoke. Very carefully bring the cross around and place the other leg into the half inserted cup, again making sure not to dislodge any of the needle bearings.

Take the next cup and start it in the opposite hole, making sure it is straight and square. Using the vice or press push it in until it almost reaches the cross. Now move the cross back so that it is half way into the just inserted cup, the other leg still half in the first cup. Check that all needle rollers are still in place.

Both the cups can now be pressed home, using the round rod to push them into the yoke (Figure 5.44). Insert the locating circlips.

Figure 5.44 *Both cups pushed into yoke and cross - properly located ready for insertion of circlips.*

Figure 5.45 *Easing a binding yoke. Do not damage seal on bearing cup*

Repeat the procedure for the other yoke and cups.

If the universal tends to bind in one direction, hold the yoke in a vice and give the end of the yoke a hit with a soft drift (Figure 5.45). This may have to be repeated on the other side of the yoke.

Shaft couplings

Couplings are used to connect two shafts on the same axis so that one can transmit power to the other.

There are several different types, the particular type used will depend on the power that has to be transmitted. Some of the more common ones include rigid couplings, flexible couplings and mechanical flexing couplings.

Rigid couplings

These may be either flanged and split sleeve couplings. These are used where there is no likelihood of misalignment or lateral movement of the shafts (Figure 5.46).

Flexible couplings

These include Flexicross and Jaw couplings. These couplings have two half bodies having two or three driving dogs, into which is assembled an oil resisting rubber cross. This rubber cross, besides transmitting the torque, protects the source of the power from shocks or impulses, and allows a small amount of angular and axial misalignment (Figure 5.47).

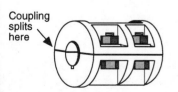

Coupling
splits
here

Figure 5.46 *Split sleeve rigid coupling*

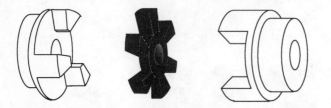

Figure 5.47 *Jaw coupling*

Mechanical flexing coupling

This coupling consists of two in-line hubs, whuch are machined with sprocket teeth, connected by a length of duplex roller chain. An oil tight cover protects the coupling and retains grease for lubrication.

Causes of coupling failure

There are many problems that can arise with flexible couplings, but proper maintenance will alleviate most of the causes of coupling failure.

- Inadequate lubrication will significantly reduce the working life of mechanical flexing couplings.
- Loose set screws or grub screws, and loose fitting keys in keyways in the shaft or the coupling hubs on the shaft will lead to failure of the coupling.
- Coupling guards should give adequate clearance and not interfere with the operation of the coupling.

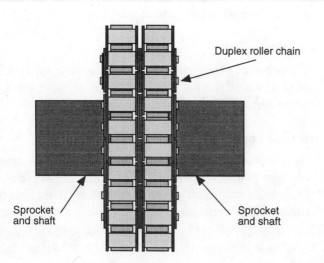

Figure 5.48 *Schematic view of a mechanical flexible chain coupling*

- Excessive build up of dirt and other contaminants around the coupling may cause it to bind, and prevent its proper operation.
- Overloading. If the operating conditions exceed the design limit of the coupling it will result in rapid wear and premature failure.

Slip clutches

Slip clutches are fitted to most power driven machinery using direct drives. They protect the drive train of the machine from damage if a malfunction occurs. The slip clutch will protect components such as the gear box, chain and sprocket drives, flexible couplings, key and keyway drives.

If the main drive on the machine is a belt drive, the belt will usually slip before damage occurs to any other components.

Machines fitted with slip clutches need special attention before each season. The clutch should be dismantled, clutch plates inspected for wear and damage, and reassembled in the reverse order. Tighten the bolts so the nuts just tension the springs. Make sure the clutch will slip by locking the machine up with a bar placed in a convenient place, and engaging the PTO. This will also assist in shining the clutch plates if they are at all rusty. Final tightening of the bolts should put just enough tension on the springs to drive the machine. Over-tightening will make the clutch inoperative. Ensure all bolts are tightened to the same tension.

Slip clutches can freeze up (not slip when required) especially if left out in the weather through the season. The freezing is usually caused by moisture getting between the clutch plates and rusting the plates together. Slipping the clutch every morning usually avoids this.

This clutch should be checked and tested regularly to see that it will slip if there is an obstruction.

Shear pins and bolts

Many machines fitted with slip clutches are also protected from shock loading damage with shear pins or bolts. If a sudden obstruction is encountered the shear bolt would shear before the slip clutch has time to slip.

The slip clutch will keep pressure on the obstruction until the drive is disengaged, which can cause further damage.

Conversely the shear bolt, once sheared, disconnects all power and no further damage is likely.

Other machines use only shear bolts for protection against shock damage. Some mouldboard and deep ripping ploughs have shear bolts in the ground engaging parts to protect them from shock loading if they encounter a rock or stump.

When replacing shear bolts or pins use only the size and grade shear bolt recommended by the particular machine manufacturer. Never use a mild steel or high tensile bolt. Mild steel will not shear cleanly; it tends to drag apart. High tensile bolts may be too strong and damage could occur before they shear off.

Safety guards

Under the *Occupational Health and Safety Act*, Section 15, an employer must provide a safe workplace for all employees. This means there must be no chance of an employee or others, getting caught in unprotected machinery parts. Heavy penalties will apply if the act is contravened.

All moving parts, chains and sprockets, belts and pulleys, open gears, shafts, universals and flexible couplings must by law have a protecting guard or shield to protect operators or bystanders from those moving parts.

If guards have to be removed for service and maintenance, they must be replaced before the machine is started up.

Figure 5.49 *A correctly fitted PTO guard*

CHAPTER 6

Fuel systems and air cleaners

Today fuel supplies are rarely stored in drums, but if it is necessary, the drums should be kept out of direct sunlight where possible. They should be kept sealed until they have to be used. When unsealed, make sure the drum is tipped slightly so any water that may collect on the top of the drum will run off before it gets to the bung or pump (Figure 6.1).

If the drums have to be moved, let the contents settle before using the fuel so any sediment or water will gravitate to the bottom and will not be pumped into the fuel system of an engine. Water takes anything up to 24 hours to settle out of distillate.

Where possible have bulk fuel storage tanks, installed in such a way that sediment and water condensation drains to the end opposite to the fuel take off. The lower end should be fitted with a valve so the water and sludge can be drained off when necessary, without having to empty the whole fuel tank.

Petrol stored in above ground bulk tanks should always be located in the shade. It is surprising how much petrol will be lost by evaporation from bulk storage tanks. It has been calculated that 25 to 30 litres for every 1000 litres stored can be lost to evaporation per month. The part of the petrol that evaporates first burns best, so the leftover petrol is harder to burn, making engines harder to start, and giving less power.

Diesel fuel systems

The fuel system consists of several different parts including the fuel tank, the lift pump, a series of filters, the transfer pump, the high pressure injector pump, the injector, and high and low pressure lines.

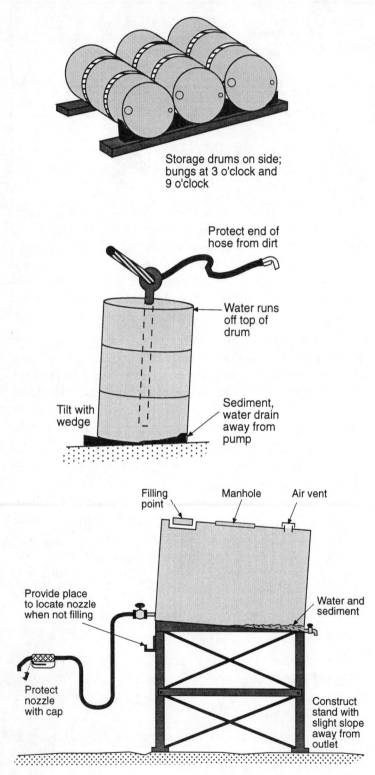

Storage drums on side;
bungs at 3 o'clock and
9 o'clock

Protect end of
hose from dirt

Water runs
off top of
drum

Tilt with
wedge

Sediment,
water drain
away from
pump

Filling
point

Manhole

Air vent

Provide place
to locate nozzle
when not filling

Water and
sediment

Protect
nozzle
with cap

Construct
stand with
slight slope
away from
outlet

Figure 6.1 *Correct fuel storage can minimise contamination*

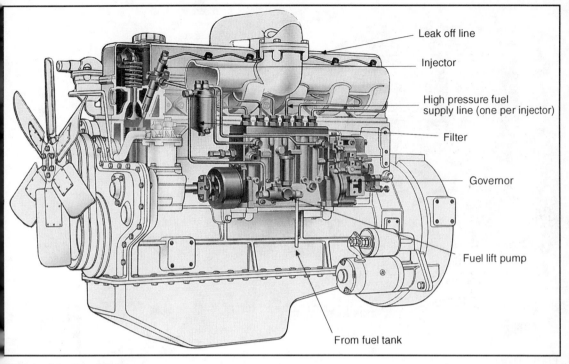

Leak off line

Injector

High pressure fuel
supply line (one per injector)

Filter

Governor

Fuel lift pump

From fuel tank

Figure 6.2 *Diesel fuel system components*

Firstly we should look at the fuel which is used in the system. All distillate contains a waxy material. In summer or warmer weather this wax remains liquid and will almost be invisible. However in winter and cooler temperatures this wax becomes a solid mass and will restrict or even completely block the fuel filters.

Distillate has a summer formula and a winter formula. If for some reason the fuel tank has been filled just before winter with a summer formula, problems can be expected with diesel engines on very cold mornings. It will usually be this waxy material solidifying in the filters, not letting enough fuel through, and starving the engine for fuel. This will make the engine very hard to start.

Lift pump or transfer pump

Fuel from the fuel tank is fed to the lift pump through a low pressure pipe. This may be gravity fed if the tank is above the pump, or if the tank is below the pump, the pump will suck the fuel from the tank.

The lift pump supplies fuel to the injector pump at a constant pressure, usually from 4–7 psi depending on the design of

1 Filter drain screws
2 Secondary fuel filter
3 Primary fuel filter
4 Filter attaching screws
5 Priming lever of transfer pump
6 Filter bleed plugs
7 Sediment bowl clamp
8 Fuel screen
9 Glass sediment bowl

Figure 6.3 *Primary filter, secondary filter lift pump and sediment bowl. (Courtesy John Deere)*

the fuel system. It has to pump this fuel through the filtering system.

Fuel filters

The fuel filter is a very important part of any fuel system. It is the first line of defence in preventing any foreign material entering the very close tolerances of the fuel system. This foreign material is usually in the form of dust, rust or water and will damage the injector pump or injector if it is allowed into these precision parts. Fuel filters should be the first thing checked when the engine loses power, or the operation of the engine begins to falter.

Most fuel systems will have a glass sediment bowl, usually part of the lift pump. This is to collect larger pieces of dirt or rust and water. It should be checked regularly and cleaned when dirt or water is visible in the bowl.

To clean a sediment bowl, close the fuel tank tap, unscrew the locating clamp and remove the bowl, taking care not to lose gasket or seal. Wash out the bowl with clean distillate and wipe out with clean lint free cloth. Replace the bowl, making sure the gasket or seal is in place before tightening. Open the fuel tank tap and let fuel fill bowl, then tighten locating clamp and check for leaks. A new gasket or seal may have to be used.

Some machines have another two filters; primary and secondary filters. They may be cartridge type, where the case

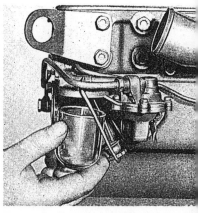

Figure 6.4 *Replacing a filter sediment bowl after the glass has been cleaned*

Figure 6.5 *Removing a disposable cartridge from a fuel filter*

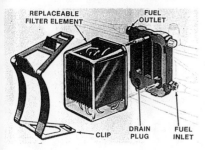

Figure 6.6 *A completely throwaway filter*

has to be removed and a new cartridge fitted, or the throw away type where the whole unit is replaced. When replacing filters, all old gaskets must be removed and replaced as the new elements are fitted.

Care must be taken when removing filters as some have small washers and springs that must be replaced in the correct order to ensure the filters function properly and do not leak.

Be sure to bleed out any air from the filters before attempting to start machine. The operator's manual should be consulted for additional information.

Injector pumps

Injector pumps are many and varied in design and appearance. There are two main types; the "in line" pump (Figure 6.2) and the "rotary" or distributor type (Figure 6.7). Both supply fuel under pressure to the injectors. This pressure varies from 1000–10 000 psi depending on the manufacturer's design.

Injector pumps are complex and sophisticated pieces of equipment and should only be serviced by a qualified mechanic. However the in line pump usually has a separate lubrication system and the oil level in this type of pump should be checked regularly.

Diesel fuel injectors

The injector is a precision part and its proper function is of utmost importance in a diesel engine. They come in a wide variety of designs, but most agricultural diesel engines use some variation of the "hydraulic pressure" principle which uses the pressure of the fuel to lift a needle valve by bringing pressure against a pre-set spring.

Most injectors are designed to let some fuel pass by the stem of the needle valve. This fuel then becomes the lubricant and the cooling medium for the metal to metal contact parts of the injector. It is finally returned via the bypass back to the fuel tank. The injector nozzle atomises the fuel and directs this spray into a critical spot in the combustion chamber to mix with the incoming air for efficient burning.

As previously mentioned the injector is a precision component and should only be tested, dismantled and cleaned by a qualified mechanic. However anyone should be able to diagnose a faulty injector and remove it.

Figure 6.7 *Rotary injector pump. (Courtesy John Deere)*

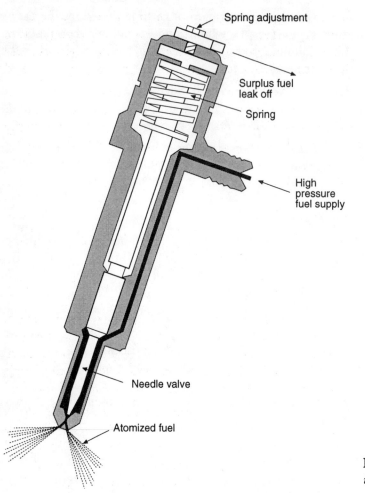

Figure 6.8 *Simplified hydraulic pressure injector*

A faulty injector may be suspected if the engine smokes without load, loses power, runs erratically or jumpily, or if the engine misfires on one cylinder. A faulty injector will show up more readily at idle speed.

Locating the faulty injector
Loosen the high pressure line to each injector in turn with the engine running at idle speed. This allows the fuel to escape before it reaches the injector and will in effect cut out that injector and the cylinder. The cylinder least affecting engine performance is usually the one with the faulty injector.

Removing the injector
The following tools are required; injector puller bar, carbon reamer, brass wire brush and a torque wrench.

First make sure the engine and fuel lines are absolutely clean. With the correct size spanner disconnect the high pressure line and the return fuel line. Cap the disconnected lines and fittings immediately with plastic caps to keep any dirt and foreign matter out of fuel system.

Remove the injector clamp nuts, and use the injector puller bar to remove injector from the cylinder head. Make sure the solid copper gasket is removed with the injector. If the gasket does not come out with the injector it must be removed from the cylinder head.

Installing the injector

The injector recess in the cylinder head must be thoroughly cleaned. This cannot be overemphasised. Pay particular attention to the sealing surface at the nozzle end. Use the carbon reamer for this task. Blow the loose carbon out of the recess by engaging the decompression lever and turning the engine over with the starter motor. If a decompressor is not fitted, make sure the fuel supply is cut off and turn the engine over in the same way. Lack of a good clean seal can result in combustion blow by, overheating or cocking of the nozzle holder that may cause the nozzle valve to stick or bind.

Make sure the old gasket has been removed and install a new copper gasket on the new or reconditioned injector. This gasket is important in positioning the injector in the head, and for the proper cooling of the injector. Insert the injector slowly, taking care not to bump the nozzle tip on anything hard. This can bend or distort the nozzle.

It should never be necessary to force the nozzle into the cylinder head. If it does not fit freely, all the carbon deposit may not have been removed.

Tighten the injector clamp nuts alternately to prevent the injector cocking in the cylinder head. Finally, using a torque wrench tighten the nuts evenly to about 22 Nm (40 lb-ft). See the operator's manual for exact details.

Before connecting the high pressure line to the injector, turn the engine over until fuel flushes the line. Now reconnect the high pressure line and the bypass bleed off line. If the engine will not start, "crack" the injector line by loosening it half a turn to bleed any air out of the line.

With the engine running check all lines for leaks, and check the injector for "blow by" past the nozzle holder. Blow by could be caused by a cocked injector, dirty nozzle cap seating surface, or damaged copper gasket. Any fault must be corrected immediately.

Bleeding diesel fuel systems

If air enters a diesel fuel system for any reason the system must be "bled" of all air. Air can get in by running out of fuel, changing fuel filters or through cracked or broken fuel lines. The quantity of air that enters the system will determine how far the bleeding process has to go.

If you run out of fuel, and hear the engine cough once, and immediately pull the stop button, you may only have to bleed the system to the injector pump. Sometimes just before the fuel runs out, the engine will inexplicably speed up a little. If the stopper is pulled before the engine stops bleeding may not be necessary. The engine might run for a minute or so after refueling and then stop. This will require bleeding of the full fuel system.

Consult the operator's manual for the correct sequence of the bleeding procedure for that particular engine. It can vary on different engines. If no operator's manual is available, follow the steps below and the engine may run.

To bleed any fuel system you will need to have fuel in the fuel tank. Loosen the bleed screw on the primary filter and pump the hand lift pump until all air and bubbles are replaced with a fine stream of fuel. Repeat the procedure on the secondary filter. Locate the bleed screw or screws on the injector pump, loosening the lower one (if more than one) first. Pump the lift pump again until all air and bubbles are replaced with a stream of fuel. Repeat for second.

To bleed the remainder of the system, (injector lines and injectors) the engine must be turned over. With the correct sized spanner loosen the connection of the injector pipe and the injector by half a turn. Now turn the engine over using the starter motor until all air and bubbles are replaced by fuel only. Tighten the injector coupling and see if the engine will start. If not, repeat the procedure on the next injector, and so on until engine fires.

If engine runs roughly there may still be air in the system, possibly in only one injector line. Follow the procedure for locating a faulty injector and if rough running persists bleed the whole system again.

Figure 6.9 *Centrifuge type pre-cleaner*

Figure 6.10 *An oil bath air cleaner*

Air cleaners

Air used in the combustion chamber must be clean. If dust or dirt is present in the air some of it will lodge on the cylinder wall to be rubbed up and down between the piston and the cylinder wall. This will increase wear and eventually the engine will have to be reconditioned.

Several types of air cleaners are used to clean this air. The first is usually a pre-cleaner. The main air cleaner can be either a dry paper cartridge, oil bath or oiled foam.

Pre-cleaners

Modern pre-cleaners work on the centrifuge principle. As the air is drawn through, it is directed in a circular motion. The heavier material (dirt and dust) being thrown to the outside by centrifugal force. Foreign matter is collected in a clear plastic bowl at the top of the air intake. This bowl should be emptied regularly to keep the heavy material from entering the main air cleaner.

Oil bath cleaners

For many years the oil bath air cleaner was the standard type of air cleaner. In operation air is drawn through a bowl of oil, up through varying layers of gauze and steel wool. The oil covers the gauze and steel wool, and any dirt particles are held by the oil. When the engine is stopped the oil and dirt drain back into the bowl, the dirt settling to the bottom. This dirt has to be cleaned out regularly by removing the bowl, tipping the oil out, and washing the dirt out with kerosene. The bowl is then dried and filled to the oil level mark.

Dry paper cartridge cleaner

Most modern engines now use a dry paper cartridge air cleaner. This type of air cleaner comes in many shapes and sizes, but all have a special paper element. This element must allow air to pass through without restriction but must stop even the smallest particle of dust from being drawn into the engine.

The element must be checked regularly for dust build up. Any dust can be removed by gently bumping the edge of the air cleaner element on a soft surface, (the front tyre is usually a suitable object. The element must be held so the dust falls away and does not get into the clean centre section of the element.

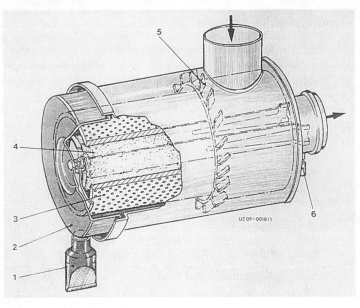

1 Dust discharge valve
2 End cover
3 Filter element
4 Safety element
5 Cyclone separator
6 Vacuum switch for maintenance
indicator

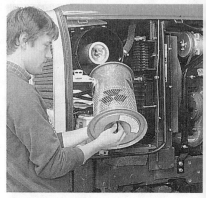

Figure 6.11 *Typical two stage dry paper cartridge element air filter. The inner safety filter is not removed and cleaned, but is replaced after a certain number of engine hours have elapsed as recommended by the manufacturer. (Courtesy of Mercedes-Benz and Massey Ferguson)*

Compressed air can also be used to remove dust. The air director valve must direct air from the inside of the element, held at least 50 mm from the paper, with maximum pressure of 100 psi. If any damage is noticed to the paper element, it must be replaced.

A special air cleaner element detergent can also be used to wash the element. Only the special detergent should be used, strictly following the directions for the product.

Oiled foam cleaners

Oiled foam air cleaners are only used on small engines, and sometimes on lawn mowers and chainsaws. They consist of a block of synthetic foam material that has been soaked in oil, with the excess oil removed. Too much oil will restrict the air flow and cause hard starting, rough running, and black smoke.

Oiled foam air filters need very regular servicing and cleaning. They have to be washed in a solvent to remove old oil and dirt, and re-oiled. Two stroke fuel can also be used to wash the foam. Oil residue from the fuel should be enough for the re-oiling of the foam.

Figure 6.12 *Cleaning a paper element type air cleaner with compressed air, directing air from inside*

> **Caution: When using volatile fuels always pay attention to fire safety precautions.**

Carburettors

The carburettor is a very important component of any petrol engine. Its task is to mix petrol and air in the correct proportions so that the mixture will burn efficiently in the engine.

Laboratory tests have proven that the best fuel air ratio is 15 parts by weight of air to 1 part by weight of petrol. It is not generally appreciated that air has weight and that this weight has a definite bearing on carburettor design, construction and adjustment.

One thousand cubic centimetres of air weighs approximately 15 grams and a thousand cubic centimetres of petrol weighs approximately 1000 grams. To put that into perspective we will need 15 cubic metres of air (that is a reasonably sized semi-trailer load) to burn one litre of petrol. (This is one reason to keep the air filter well serviced. Imagine the amount of dirt that could get into an engine in dusty conditions if the engine were burning 15 litres of fuel per hour.)

Many carburettors have few external adjustments. Changing the fuel mixture requires the size of the main jet to be changed. Changing the float level involves removing the carburettor from the motor, and taking the top off to expose the float.

Figures 6.13 to 6.16 give some idea of the workings of a simple up draft carburettor as fitted to some small petrol engines.

Float

The float controls the level and supply of petrol in the fuel bowl. When the fuel bowl (1) is empty, the float and lever (2) and float valve (needle) (3) drop (see Figures 6.13–16). Petrol under pressure from the petrol pump (or gravity feed from the petrol tank) is forced through the float valve seat (4) into the fuel bowl. As the bowl fills the float and lever will rise pushing the float valve into the seat and stopping the flow of petrol into the bowl. As petrol is used through the carburettor the float will fall, allowing more petrol to enter the bowl.

Under actual working conditions the fuel level (5) and the float and lever (2) will maintain a position so that the inward flow of petrol is equal to the outward flow of petrol to the engine.

If the float valve and float valve seat (needle and seat) wear, or if dirt or other matter lodges on the seat preventing the float valve from closing, an oversupply of petrol can occur in the fuel bowl, resulting in the engine flooding (getting too

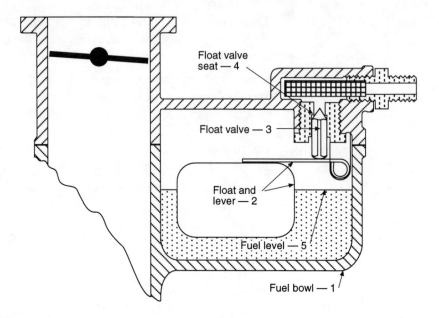

Float valve seat — 4

Float valve — 3

Float and lever — 2

Fuel level — 5

Fuel bowl — 1

Figure 6.13 *Float system*

much petrol). This will cause the engine to run roughly, lose power, blow black smoke and generally sound sick.

Idle system

The idle system controls the flow of fuel at idle speed and at slow speeds until the throttle is opened wide enough to allow the power fuel system to function.

When the throttle valve (6) is in the idle position, the edge of the valve is between the primary idle orifice (7) and the secondary idle orifice (8). With the valve in this position the air pressure (manifold vacuum) at the primary idle orifice is lower than air in the air pressure in the fuel bowl chamber (9) and fuel is forced from the fuel bowl (1) into the idle fuel passage (10). As the fuel passes through the idle fuel passage (10) it passes through the metering orifice of the idle jet (11) to a point where it is combined with air entering through the idle adjusting needle seat (12).

The mixing of air with petrol helps to atomise the fuel. As this rich mixture of petrol and air emerges from the primary orifice (7) it is reduced to the correct proportions by the air which passes around the throttle valve (6) provided that the

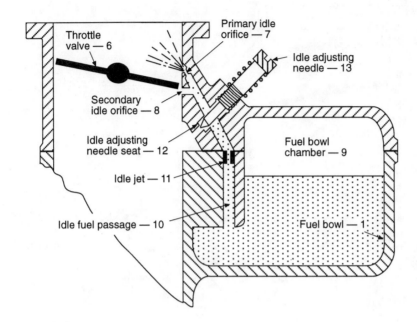

Figure 6.14 *Idle system*

idle adjustment needle (13) is properly adjusted. The throttle valve must be open a small amount to allow the engine to idle.

If at any time these small holes become blocked with dirt or other matter, the idling of the engine will be adversely affected. These holes will need to be cleaned. Never use wire or drills as these will change the shape and size of the holes upsetting the whole calibration of the carburettor. Use only a cleaning fluid and/or compressed air for cleaning.

Power system

With the throttle valve (6) in the slow or just off slow idle position, petrol rises up through the nozzle (14) and out of the nozzle air bleeds (15) to fill the accelerating well (16) to approximately the height of the fuel level in the bowl.

As the engine speed increases, the air flow also increases, until the velocity through the venturi (17) is high enough to create a pressure at the tip of the nozzle (14) slightly less than the pressure in the fuel bowl. Fuel therefore feeds from the fuel bowl through the opening between the power adjusting needle (18) and the power adjusting needle seat (19) through

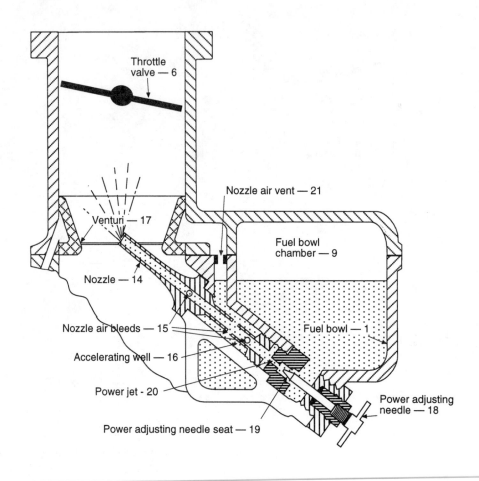

Figure 6.15 *Power system*

the power or main jet (20) and out the nozzle (14) to be discharged into the air stream at the venturi.

The amount of petrol being used by the engine can be controlled by the adjustment of the power adjusting needle (18).

Too much petrol will make the engine run roughly, as in a flooding engine. Flooding will also make the engine very hard to start. Too little petrol will result in loss of power. The power adjusting needle should be adjusted until the engine runs smoothly under load.

Choke system
The function of the choke valve (25) is to restrict the amount of air that can enter the carburettor and increase the suction

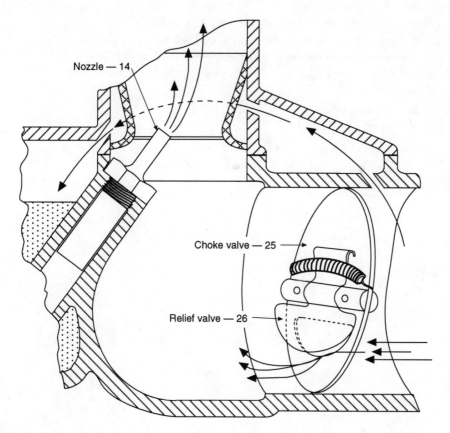

Nozzle — 14

Choke valve — 25

Relief valve — 26

Figure 6.16 *Choke system*

on the nozzle (14) so that additional fuel will be drawn into the engine.

As soon as the engine fires and runs, the rich mixture must be rapidly reduced to prevent stalling. This change in mixture is accomplished by the operator positioning the choke valve to provide the proper mix. A few degrees movement of the choke valve will make a big change in the mixture strength.

To help reduce the sensitivity of the choke valve position, a spring loaded relief valve (26) is used in many carburettors. This valve opens automatically as the engine increases in speed and eliminates a great deal of the choke manipulation required by the operator.

If the motor runs roughly and black smoke is expelled from the exhaust, check:

• that the choke is fully open and not stuck in the closed or half closed position,

- that the air cleaner is not blocked,
- that the power adjusting screw (mixture screw) has not worked loose. Turn the screw in several times and see if any difference is made to the running of the engine. If so, adjust screw until the engine runs at its best, and tighten lock nut (if fitted).

If petrol runs out of the carburettor after the engine has stopped it could indicate:

- that the float level is too high,
- that the float valve and float valve seat are leaking; due to wear or something fouling the seat.

CHAPTER 7

Hydraulic systems

When we refer to agricultural hydraulic systems we usually refer to either three-point linkage, using an internal cylinder, or remote hydraulics using an external cylinder. Remote cylinders are used on drawbar type implements like the trailing plough, combine, mower or header. Here the cylinders lift implements out of the ground, or raise and lower them. The external cylinders may be used on an implement fitted on the tractor such as front end loader or fork lift.

Self-propelled headers and windrowers have their own hydraulic system for raising and lowering various components.

Figure 7.1 *Components of a three-point linkage*

Rockshaft

Control levers

Right lower linkage adjustment

Top link pins and linch pins. Note the use of chain to ensure that they are not lost

Top link adjustment

Adjustable stabiliser bar

Left lower link end

Right lower link end

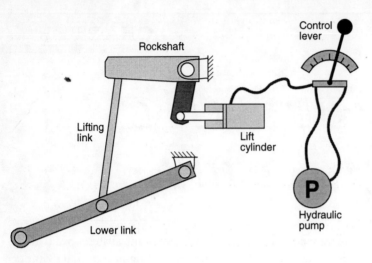

Figure 7.2 *Schematic representation of a three-point linkage in operation. The control lever operates a valve in the hydraulic system, causing the lift cylinder to expand (or contract), which, in turn, causes the lower link to be lifted (or to be lowered)*

Hydraulic components

The hydraulic system includes the following components:

- hydraulic cylinder,
- control valve,
- hoses,
- breakaway couplings.

Hydraulic cylinders (Rams)

Hydraulic cylinders come in all shapes and sizes depending on the task performed. They are usually classified by:

- diameter of the cylinder,
- length of stroke of the piston,
- action, single or double:
  - single acting exert pressure in one direction only, they rely on gravity or spring tension to return the piston for the next cycle.
  - double acting exert pressure in both directions.

Why do hydraulic cylinders sag?

Several things can cause a hydraulic cylinder to sag. The most likely cause is the piston seal or O-ring is worn or damaged and it allows the oil past it. This allows the piston to move in the cylinder without operating the control levers.

Seals do eventually wear, but this wear can be accelerated if there is dirt in the hydraulic oil. Dirt will tend to stick to the seals and O-ring and be dragged backwards and forwards along the cylinder bore, wearing both the cylinder and the O-ring.

Figure 7.3 *A typical agricultural double acting ram*

The control valve might be worn. This also lets oil past without operating the controls. Wear is caused again by dirt in the oil.

There may be some foreign object in the control valve preventing it from closing completely. Objects may include thread tape that has worked its way free from an incorrectly wrapped connection or joint. It is not recommended to use thread tape on any hydraulic fitting or connection.

Pieces of rubber or steel hose braid, entering the hydraulic system when fitting new ends on hoses, may be holding the control valve open.

Lint from a rag used to wipe the oil dipstick can prevent the control valve from closing. Always use lint free rags.

Part	Other common names
Cylinder	barrel, tube
Rod	shaft, piston rod
Gland	gland nut, gland plug, end cap, rod guide
Piston	piston head
Gland seal groove	rod seal, shaft seal
Piston seal groove	piston seal head
Rod wiper groove	wiper, scraper, dust seal
Port	inlet and outlet holes
Bush	bearing, sleeve
End cap	anchor end

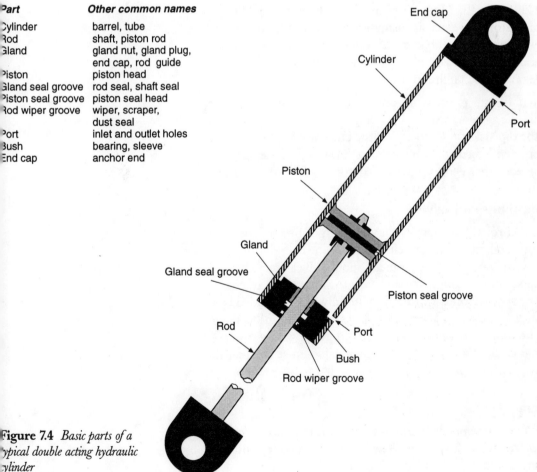

Figure 7.4 *Basic parts of a typical double acting hydraulic cylinder*

This problem can sometimes be rectified by holding the control valve wide open and revving the motor to pump maximum oil through the valve. It may carry the foreign body out of the control valve. The hydraulic oil filter, if fitted, may pick up the foreign body before it does permanent damage.

The cylinder rod or shaft seal may be leaking. This can be seen if an inspection of the cylinder is made. Oil will be found around the rod seal or wiper seal.

There could be a leaking hose that has gone unnoticed because of its inaccessibility. It can be detected from oil dripping from the machine or a pool of oil under the machine when left standing.

> **Note: Only qualified mechanics should attempt to dismantle a control valve.**

Hydraulic reservoir

This is the oil storage tank. On most tractors this storage is also the lubrication storage for the transmission (gear box and differential). Most other machines have a separate hydraulic storage reservoir, using special hydraulic oil. Every effort must be made to keep dirt out of these reservoirs as the tolerances in the hydraulic system are very fine and any dirt will soon cause failure in the components.

When filling a reservoir on a machine with a separate hydraulic system, it should never be filled over the full mark on the dipstick (or 2/3 to 3/4) of the tank capacity. This space is for what is termed "displacement variation" of the oil being pumped to the cylinder and the oil returning from the cylinder.

Control or spool valve

The control valve controls the direction of flow of the high pressure oil, directing it to either end of the cylinder depending on the task to be performed. Spool valves are so named because the working part is similar to some types of cotton reel or spool. Control valves should only be dismantled by a qualified mechanic. There are many small parts and damage can be done if these parts are replaced in the wrong order. If a fault does occur the whole control valve must be taken to an expert for repair.

Hydraulic pump

This pump can be of several different types like a gear pump, piston pump or vane pump. The flow rate or litres per minute pumped varies considerably with different machines, but it

Figure 7.5 *Cross section of a spool valve. Operating the control lever moves the spool in and out, opening or closing the various ports to direct the oil as required*

will have a great effect on the speed of travel of the cylinders. This in turn affects the speed of operation of hydraulic attachments. All pumps will have the capability to pump to a pressure of between 2000 and 3000 psi or approx 14 000 to 21 000 kPa. They should be fitted with a pressure control valve to protect the hydraulic system if it becomes over pressurised. This usually happens when the cylinders reach the full extremity of their travel in either direction on double acting cylinders. Only authorised mechanics should adjust these pressure control valves.

Hydraulic hose

The rubber single braid reinforced hose is made up of:

- an inner core of synthetic seamless oil resistant rubber hose,

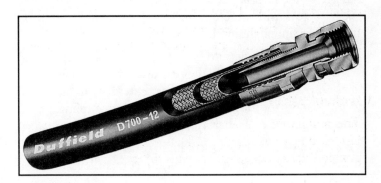

Figure 7.6 *Double braid hydraulic hose. Note: Non-skived, re-usable hose end. (Courtesy Duffield)*

- a covering of braided high tensile steel wire,
- an outer cover made of synthetic rubber resistant to sunlight, oil, weather and abrasion.

Double braided hose is made the same way except two braided high tensile layers are used, separated by another layer of rubber.

Breakaway couplings

The hydraulic couplings connecting a remote cylinder on a drawbar implement to a tractor are usually called breakaway couplings. They make hitching and unhitching much simpler and quicker than would be the case if the hydraulic hoses had to be unscrewed.

The couplings are made with "male" and "female" ends, the female end usually on the tractor. The term breakaway was probably first used because the couplings were supposed to come apart if the drawbar pin came out and left the machine behind, thus avoiding damage to the hydraulic hoses. However in practice this does not always happen.

There are many sizes and styles of breakaway couplings. Most tractor manufacturers do not use common breakaway couplings. However, they can be changed so one coupling fits all makes of tractors.

Caution: Tractor manufacturers often specify their own, or a particular brand or grade of oil in the transmission and hydraulic system. Any other oil may cause severe damage to the system.

If different makes of tractors are used on the same hydraulic cylinder the oil in the cylinder from the first tractor will contaminate the oil in the second system. This can cause severe damage to the hydraulics and transmission. The practice of using tractors with different oil in the hydraulic system on the same cylinder should be avoided and all times.

All hydraulic couplings and fittings should be kept clean and free from dirt. If dirt enters the hydraulic system in any way, particularly when connecting hoses, extensive damage can occur. Always thoroughly clean breakaway couplings, both male and female ends, before connecting.

It is safest and most convenient if the controls always operate in the same direction, that is, pull the control lever to raise

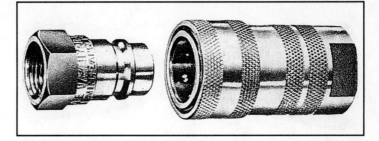

Figure 7.7 *Breakaway couplings, male and female ends. (Courtesy Duffield)*

the implement and push the lever to lower. This is particularly so if operating a loader fitted to the tractor.

It is advisable to label the breakaway couplings in some way so they can be connected the same way each time an implement is connected. Coloured insulation tape wrapped around the end of the hose is a common way of labelling the hoses.

Hose ends

There are many different types of hydraulic hose ends available, depending on the type of hose they are to fit, and the connection they have to make. One of the main differences that will concern you in maintaining machinery is some are re-usable and others are non re-usable.

The non re-usable end couplings are swaged, or crimped on with a special machine. While they are probably a more permanent hose end, if the hose is damaged in any way the whole hose and ends must be replaced. Sometimes this may mean a trip to town just to replace a hose. The ends are the most expensive part of the complete hose so it is simple to see if you can replace the hose using existing ends the job must be cheaper.

Re-usable end couplings come in two types:

1. Skive type fittings require the outer layer of rubber to be removed, exposing the wire braid. The outer fitting is screwed over this braid and the inner fitting is then screwed into this assembly.
2. Non-skive fitting is the same as above except the outer layer of rubber is left and the fitting screwed over this rubber.

If there are a number of hydraulic systems with re-usable ends on the farm, it is advisable to have at least one length of the correct diameter hydraulic hose on hand. A damaged hose can then be replaced with the least amount of down time.

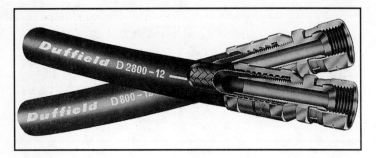

Figure 7.8 *Single braid hose, front is non-skive re-usable hose end, rear is skived re-usable end. (Courtesy Duffield)*

To fit re-usable ends to hose make sure you have the right fitting; hose diameter matching fitting diameter, fitting threads match, etc. (Note: some hose and hydraulic fitting manufacturers use code matched fittings.) Do not mix different makes of fittings on the same end of a hose.

Fitting procedure

1. Cut replacement hose to length. Use a fine tooth hacksaw or angle grinder. A hacksaw tends to leave the braid feathered. When working with an angle grinder use a metal cutting disc and wear safety glasses. If none of these implements are available, a sharp axe and a block of hardwood can be used. Make sure the pipe is cut with one clean hit.

   Clean out the end of the cut hose of any rubber or braid fragments. With a sharp knife put a slight bevel or taper on the inside and outside of the rubber end of the hose.

   Select appropriate re-usable end for hose paying particular attention to inside diameter of hose and fitting. Some fittings are code matched.

2. Determine if re-usable fitting is skive or non-skive. Some manufacturers use a marking system on the body of the fitting to indicate skive or non-skive. If unsure the taper on the inside of the skived body is usually more pronounced. Non-skive are the easiest to fit. If using a skive fitting the outside layer of rubber will have to be removed to the correct distance. A special tool is available for this procedure, but a sharp knife can be used to do the job when it is only done occasionally.

3. Fit hose end body. The procedure here will depend on the length of assembly.

   When making up long assemblies lay the hose horizontally in a vice with about 75 mm protruding. (Take care not to crush or distort the hose when gripping in the vice.)

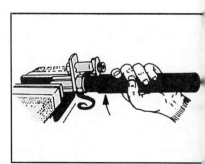

Figure 7.9 *Hose peeling device for skive fittings. (Courtesy Duffield)*

Figure 7.10 *Fitting hose end body. (Courtesy Duffield)*

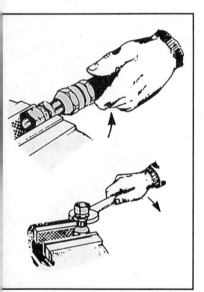

Figure 7.11 *Fitting stem. (Courtesy Duffield)*

Lightly oil the outside of the hose and screw the hose end body onto the hose by hand, anti-clockwise, until the spiral form in the body bites into the outer rubber cover of the hose.

Take a spanner, and while steadying the hose end body turn it anti-clockwise until it bottoms against the end of the hose. Back off the body by turning it clockwise 2/3 of a turn.

When making up short assemblies grip the hose end body in the vice so that the vice jaws bear directly onto the hexagonal flats.

Lightly oil the outside of the hose and screw the hose in anti-clockwise.

Firm resistance will be felt as the body bites into the outer cover of the hose. When the hose bottoms, back off 2/3 of a turn.

> **Note: some of the older type re-usable ends are not the screw on type and can be quite difficult to re-fit.**

4. Fit stem.

   Grip the stem in the vice to start it into the body.

   Oil the stem taper and bore of the hose and engage two or three threads by hand, taking care that the threads are not crossed.

   Grip the hose end body in the vice, vertically for short hoses, horizontally for long, then screw the stem home with a spanner. If excessive resistance is felt, due to hose tolerance, the stem collar can be left up to 1.5 mm from the body. The hose end assembly is now complete.

5. Clean and check.

   After assembling both hose end fittings, blow compressed air through the hose assembly to ensure the bore is clean. If slivers of rubber are left, a slight whistling sound can be heard when cleaning with compressed air.

Assembly tips

If hose tends to flare when fitting into the hose end body, cut or grind a taper or bevel on the outer rubber cover. Fit one side of the hose under the body skirt and work the rest of the outer hose cover under the body skirt with a screw driver.

If the stem is difficult to start in the body, back off the hose 2–2.5 mm from the fine thread in the body.

Care should be taken during stem entry to ensure that the leading edge of the stem does not shave off small slivers of rubber from the inner hose which could find their way into the hydraulic system.

Thread tape is not recommended for use in hydraulic fittings. If thread tape is used on threads, care must be taken that tape will not enter the hydraulic system. Damage will result to the control valves and pump.

Hose assembly routing tips

Proper hose installation is essential if long term hydraulic hose problems are to be prevented. The following diagrams show the right and wrong way to install hoses in a variety of situations, using various types of fittings.

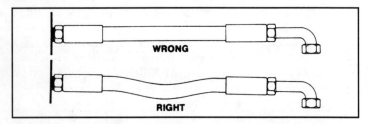

When hose installation is straight, allow enough slack in the hose to provide for the length change which will occur when pressure is applied.

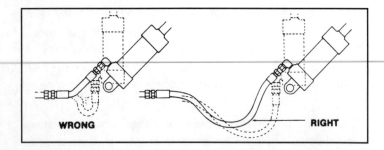

Adequate hose length is necessary to distribute movement on flexing applications and to avoid abrasion. In this layout, when looking down from above, the hose and both ends should be in the one straight line to avoid torsional loads.

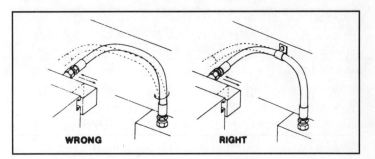

Avoid twisting of hose lines bent in two planes by clamping hose at the change of plane.

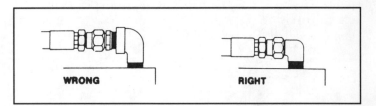

Reduce the number of pipe thread joints by using proper hydraulic adaptors instead of pipe fittings.

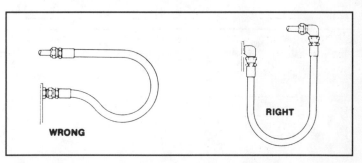

When hose radius is below required minimum, use an angle adaptor to avoid sharp bends. (Note: different grades of hydraulic hose have different minimum radii.)

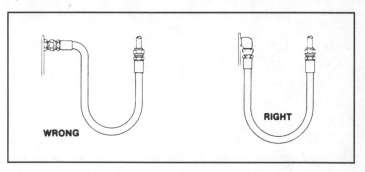

Use proper adaptors to avoid sharp twists or bends in a hose.

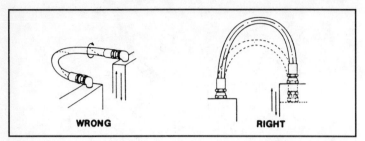

Prevent twisting and distortion by bending the hose in the same plane as the motion of port to which the hose is connected.

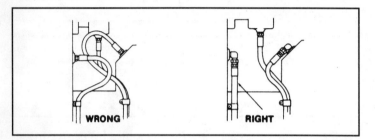

Route the hose directly by using 45° and/or 90° adaptors and fittings. Avoid excessive hose length, to improve appearance.

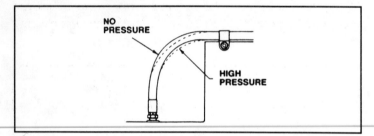

To allow for length changes when hose is pressurised, do not clamp at bends; curves should absorb changes. Do not clamp high and low pressure lines together.

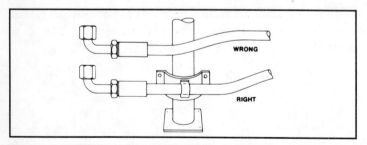

High temperatures shorten hose life. Make sure the hose is kept away from hot parts. If this is not possible, insulate the hose.

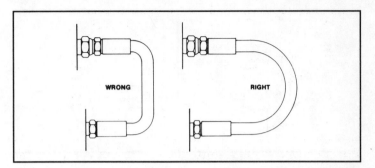

To avoid hose collapse and flow restriction, keep hose bend radii as large as possible. Refer to hose specification tables for minimum bend radii.

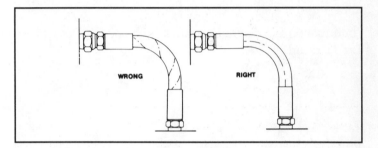

When installing a hose, make sure it is not twisted. Pressure applied to a twisted hose can result in hose failure or loosening connections.

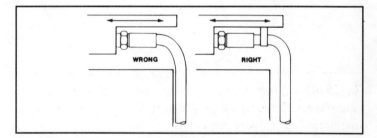

Elbows and adaptors should be used to relieve strain on the assembly, and provide neater installations which will be more accessible for inspection and maintenance.

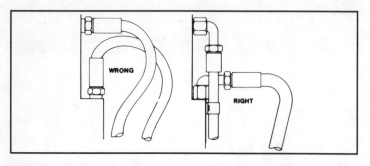

Run hose in the installation so as to avoid rubbing and abrasion. Often clamps are required to support long hose runs or keep hose away from moving parts. Use clamps of the correct size. A clamp too large allows the hose to move and cause abrasion to the hose under the clamp.

(Information and diagrams courtesy of Frederick Duffield Pty Ltd Hydraulics.)

CHAPTER 8

Cooling systems

Water cooling systems

Most liquid cooling systems fitted to farm machines and vehicles are of the pressurised type, (pressurised 4–8 psi which raises the boiling point of the coolant) and consist of radiator, hoses, water pump, fan and thermostat. All components need regular checking.

General inspection of the cooling system.

1. Check cooling water level, and check for evidence of rust or scale in the water. Water in the cooling system should have a recognised rust inhibitor added in the recommended proportion to stop rust build up in the radiator and engine block. In sub-zero temperatures a suitable anti-freeze should be added in the right proportions. Engine manufacturer's recommendations for radiator additives should be strictly adhered to. Some of these additives are very detrimental to some engines, corroding cylinder liners and eating away O-rings and damaging some water pumps.

2. Remove dirt, insects, grass seeds and so on from radiator core. Use compressed air or a hose and water. Always blow out in the opposite direction of the air flow; if you blow the same direction as air flow some of the foreign matter will be driven further into the cores. Do not use high pressure cleaners to blow out radiator cores, it may damage cores or bend the fins and restrict the air flow.

3. Examine all hoses for signs of external cracks or perishing. If the hose feels spongy it is a sign of perishing from the inside and should be replaced to avoid a burst hose in the field. Make sure all hose clamps are tight and there is no sign of leaks or seepage.

4. Check both the radiator and block drain cocks for leakage.

5. Check fan belt for correct tension and wear or cracks. Cracks usually appear on the inside of the belt and as they become worse there is a chance the belt may break.

To tighten fan belt tension, loosen the adjusting stud and the other two studs on the alternator and gently move the alternator to tighten the belt. The tension on the belt varies according to the size of the belt, a thick wide belt has more surface area in contact with the pulley so does not need to be as tight as a thin narrow belt. Thin narrow belts are the most popular type of belt in use today.

Water pump

The cooling water is circulated by means of a centrifugal type water pump which is usually mounted onto the front of the engine block. It assists in the circulation of the cooling water, drawing cool water from the bottom of the radiator and pumping it through the engine block, back up through the thermostat and into the radiator for another circuit.

Figure 8.1 *A pressurised cooling system, showing the flow of coolant*

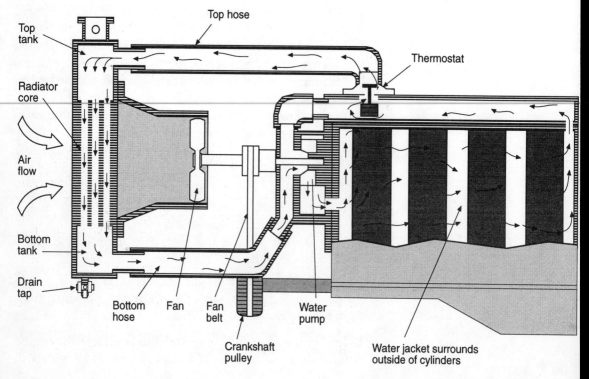

The pump is driven by the fan belt from the crankshaft which also usually drives the alternator. The fan is usually bolted to the pulley on the water pump.

The pump shaft runs on specially lubricated bearings. The pump housing has a seal between the impeller and the bearing, and if this seal leaks, water usually damages the bearing. If water is seen to seep from behind the water pump pulley it is a fair indication the seal is damaged or worn, and should be re-newed.

The water pump bearings can be tested by grasping the blades of the fan and checking it for lateral movement. If movement is detected the bearings and water pump seal should both be replaced.

To remove pump, drain radiator, remove bolts holding fan onto pulley, (the radiator may have to be removed to gain access to these bolts) and remove all locating bolts holding water pump. The water pump should now come away from the engine block. Care must be taken to remove the unit completely; the old gasket from the engine block and the water pump or any part of the gasket left will foul the new gasket and it will fail to seal properly when reassembling.

A pump repair kit usually includes new seal and bearing. Sometimes an exchange pump can be supplied ready to fit. The pulley and the impeller usually have to be pressed off the shaft, and they have to be pressed on again in the reassembly operation. If a press is not available it may be more convenient to get an exchange water pump or take it to an authorised mechanic.

Thermostat

The thermostat is a device located in the special housing bolted to the front of the cylinder head. Its job is to help control the temperature of the cooling water. As the water in the engine warms up, the hot water rises towards the thermostat. While the water is cold the thermostat remains closed and restricts the water circulation until a predetermined temperature is reached. When this temperature, 80° or 90°C, is reached the thermostat will open allowing the hot water to be pumped to the radiator top tank. This hot water is replaced by cool water coming in via the bottom hose into the engine block, so cooling the engine.

If the thermostat malfunctions and remains closed the engine will overheat. Conversely if it doesn't close the engine will run too cold or take a long time to warm up. If a temperature gauge is fitted the operating temperature should be reached in five to ten minutes. If this is not the case, and the engine is very slow to warm up it may be because the thermostat is stuck open.

To test the thermostat it must be removed. Drain the radiator, remove the top radiator hose, the two or three studs in the thermostat housing, and the housing. The thermostat will sometimes come away with the housing, or it may stay with the engine block. Care must be taken to remove all gasket material from the housing and block so the new gasket will seat properly. When the thermostat is removed, inspect for damage or corrosion. With the valve closed, place the thermostat in hot (90° C) water and see if the valve opens. If it doesn't open try it in boiling water. If it still will not open, the valve must be faulty and should be replaced with a new one.

Reassemble in reverse order to disassembly, taking care to put the thermostat in the right way. They usually have an arrow to indicate the direction of water flow. Remember that the water flows from the engine to the top of the radiator. Fit a new gasket and tighten studs evenly. Fill radiator and check for leaks.

Coolants

Water is probably the best known coolant, but in an engine it tends to go rusty after a time, unless some kind of rust inhibitor has been added. Rust inhibitor in some cases will also act as an anti-freeze.

Care must be taken when mixing additives. Read and follow directions. Only use additives recommended by the machine or engine manufacturer. Extensive damage can be caused to the engine if other additives are used.

Engine coolant should be drained and the cooling system flushed with clean water at least once per year. If it is possible back flush (the water should flow in the opposite direction to normal operation) and use compressed air, and flushing will be much more complete. The bubbling action of the water will help dislodge sludge and scale.

To back flush remove the bottom radiator hose. Remove thermostat so as not to restrict the flow of water through the system. Wrap some rag around the end of an ordinary garden hose so it will fit neatly into the bottom radiator pipe. Turn on water and let it flow through the system until it runs out of the engine by way of the bottom radiator hose. The radiator cap should be on during this operation. If an air compressor with an air director is available, air can be introduced into the water by drilling a hole, big enough to take the tip of the air director, into the garden hose, near the bottom radiator pipe. Give short bursts of air. If any of the system is rusty or corroded this procedure may cause the rust or corrosion to be removed and the system may leak. Care should be taken not to over pressurise the system.

Air cooled engines

Most small single cylinder engines are air cooled. Some larger engines and some tractor engines are also air cooled. Most Ag bikes are air cooled and rely on the large cooling fins to carry the engine heat to the air. They rely on air being pushed by a fan around or along a cowling of some kind and directed across the cooling fins.

Two things must be checked regularly on air cooled engines:

- The inlet to the fan is usually protected by a screen. When working in dusty conditions where grass, grass seeds, straw, thistles and so on are in the air, the screen can become blocked with this material. This will restrict the flow of cooling air and the engine will overheat. Check regularly that the screen is not blocked.
- If the fins become dirty they will loose their efficiency by not allowing the heat to escape. The effect of dirt and dust can be compounded if oil or grease is present. The oil may simply have been spilt onto the fins, or it may have come from faulty engine gaskets leaking oil. Whichever way, it will attract and hold dust. When the engine

Figure 8.2 *Air cooled Deutz engine; note the cooling fins on the cylinder barrels*

heats up the oil tends to burn off leaving an even thicker layer of caked on dirt, reducing the air cooling efficiency even more and resulting in a very overheated engine.

Air conditioning units

Many farm machines are now equipped with air conditioned cabs. These units provide operator comfort not even dreamed of a few years ago. Air conditioning units should give many years of trouble free comfort, provided a few basic rules are followed. They should be maintained regularly and correctly, if not the unit could soon fail.

Condenser cores

The condenser cores are usually located near or in front of the engine radiator. They should be regularly checked. Make sure the cores are quite clear of all material that will restrict the flow of air through them. Condenser cores on header air conditioning units are particularly prone to blockage because of the dusty conditions in which they are required to work. If the refrigerant becomes overheated it will not cool the cabin properly, the compressor will run continuously and further overheating and damage to the air conditioning unit could result.

The condenser cores should be cleaned with compressed air, under 60 psi or low pressure water jet. Never use high pressure cleaners, they can bend or damage the cooling fins.

Solenoid switches

From time to time check the solenoid switch. When activated it should engage the pulley on the front of the compressor. With the engine going, switch the air conditioner on. You should be able to hear the switch click in. Leave it operating for some time and switch the air conditioner off. Another click should be heard. If the switch malfunctions and does not switch off, the compressor will be operating continuously and damage to the unit could occur.

Correct belt tension is essential. A loose belt decreases the unit's efficiency and causes the belt and pulley to wear. Extra heat will also be generated by the slipping belt.

If the air conditioning system is not cooling correctly:
- Check belt tension.
- Check condenser cores.

- Check solenoid switch on compressor pulley.
- Check refrigerant. For units fitted with a sight glass, run the engine at operating revs, switch blower and air conditioner to maximum operating position, and observe the sight glass. If bubbles do not disappear within a few minutes of switching on the air conditioning system it may be low on refrigerant. You will need to call in an air conditioner expert.

Out of season maintenance

To maintain the system in good working order and to minimise the loss of refrigerant, the air conditioner should be operated for a minimum of five minutes each week regardless of the season. This will assist in preventing the compressor seal from drying out. If the seal does dry out it can cause loss of refrigerant and possible damage to the compressor unit.

This procedure is particularly important in seasonal machines like headers, cotton pickers and any machine that is put away until next season. The machine should be started up and the air conditioning system run for some time each week.

References

Dayco, *Service Manual for industrial V-belt Drives* (1994)

Fredrick Duffield Pty Ltd, Duffield Hydraulics, *Reusable System for Wire Braid Hose* (1993)

Hallite Seals International, *Sealing Systems* (1995)

Jeffrey, *Principles of Machine Maintenance*

Kepner, Bainer and Barger, *Principles of Farm Machinery*, AVI Publishing

Klinger, *Industrial Gaskets Manual* (1993)

Smith and Wilkes, *Farm Machinery and Equipment*, McGraw-Hill

John Deere, Massey Ferguson and McPhersons technical manuals.

Acknowledgments

The author and the publisher wish to thank the following for their help.

John Rockeridge of Hallite Australia for illustrations and technical information on seals.

Keith Lowe of Duffield Hydraulics for illustrationns of hydraulic fittings.

Index